Transmission
Network
Protection

POWER ENGINEERING

Series Editor

H. Lee Willis

*ABB Power T&D Company Inc.
Cary, North Carolina*

1. Power Distribution Planning Reference Book, *H. Lee Willis*
2. Transmission Network Protection: Theory and Practice, *Y. G. Paithankar*
3. Electrical Insulation in Power Systems, *N. H. Malik, A. A. Al-Arainy, and M. I. Qureshi*
4. Electrical Power Equipment Maintenance and Testing, *Paul Gill*
5. Protective Relaying: Principles and Applications, Second Edition, *J. Lewis Blackburn*

ADDITIONAL VOLUMES IN PREPARATION

Transmission Network Protection

Theory and Practice

Y. G. Paithankar
*Visvesvaraya Regional College of Engineering
Nagpur, India*

MARCEL DEKKER, INC.   NEW YORK · BASEL · HONG KONG

Library of Congress Cataloging-in-Publication Data

Paithankar, Y. G. (Yeshwant G.)
  Transmission network protection: theory and practice/Y.G. Paithankar
    p. cm.—(Power engineering; 2)
  Includes bibliographical references and index.
  ISBN 0-8247-9911-9 (hc)
  1. Electric power systems—Protection. I. Title. II. Series.
TK1010.P35 1997
621.319'2—dc21
                                                           97-36224
                                                               CIP

The publisher offers discounts on this book when ordered in bulk quantities. For more information, write to Special Sales/Professional Marketing at the address below.

This book is printed on acid-free paper.

Copyright © 1998 by MARCEL DEKKER, INC. All Rights Reserved.

Neither this book nor any part may be reproduced or transmitted in any form or by any means, electronic or mechanical, including photocopying, microfilming, and recording, or by any information storage and retrieval system, without permission in writing from the publisher.

MARCEL DEKKER, INC.
270 Madison Avenue, New York, New York 10016
http://www.dekker.com

Current printing (last digit):
10 9 8 7 6 5 4 3 2 1

PRINTED IN THE UNITED STATES OF AMERICA

Series Introduction

Power engineering is the oldest and most traditional of the various areas within electrical engineering, yet no other facet of modern technology is currently undergoing a more dramatic revolution in both technology and industry structure. This addition to the Power Engineering Series addresses a cornerstone of this revolution: secure interconnected operation of the transmission grid. Traditionally, high-voltage transmission networks provided electric utilities with additional margins of reliability and economy, worth billions of dollars annually. But as the electric power industry worldwide moves toward deregulation and competition, the interconnected transmission system becomes even more important, because it *is* the competitive marketplace; the means through which buyers and sellers in a free electric market can transact business. Without it, the electric system will not function, either electrically or economically.

This book provides a wonderfully organized and cogent view of modern transmission network protection theory and application. At both the introductory and advanced levels, it provides an above-average level of insight into the philosophies and limitations of network protection and shows a rich understanding of the structure often hidden by nomenclature and formula. In particular, readers will find Chapter 2 a very astute introduction to high-voltage protection. At the other end of the scale (and the book), Chapter 12 provides an extremely perceptive look at the future of protection relaying and some of the challenges, and progress, that can be expected in the new century.

As the editor of the Power Engineering Series, I am proud to include *Transmission Network Protection* among this important group of books. Like all the volumes planned for the series, this book provides modern power technology in a context of proven practical application and is useful as a reference book as well as for self-study and advanced classroom use. The series will eventually include books covering the entire field of power engineering, in all its specialties and sub-genres, all aimed at providing practicing power engineers with the knowledge and techniques they need to meet the electric industry's challenges in the 21st Century.

H. Lee Willis

Preface

Electrical power systems are becoming increasingly complex in nature due to integration of electrical power grids. This is a reflection of the increasing dependence of modern society on electrical energy, so much so that a short interruption in electrical supply can lead to catastrophy. Life support equipment, continuous process industries, computer networks, and communications all demand uninterrupted supply.

Even though power protection engineers are under immense pressure to keep their protection systems on standby at every moment, faults can occur. This requires the utmost reliability and accuracy of protective systems.

Protective relays have evolved from electromechanical instruments with trip contacts to the present processor-based numerical relays. The operation of the earlier relays was based on interaction between two sinusoidal fluxes to produce operating and restraining torque.

Static or analog relays, however, are based on comparison of magnitude and/or phase of the relaying quantities. Processor-based relays incorporate the additional dimension of numerical computation and logical decision making. Presently, relays that also make use of nonfunctional quantities are under development.

There is another far-reaching change, with respect to the relaying, that has slowly but surely established itself. The potential of computer-based relays (i.e., numerical relays) can be fully exploited when they are endowed with communication ability. This confluence of computing and communication generates a synergy of its own and gives modern relays a power far beyond what was originally envisaged by the designers of computerized relays.

Because we have no control over the deionizing time of arcs, it is debatable whether we should continue to strive for faster and faster relays. Perhaps we should strive for accuracy and thus ensure discrimination. These factors make this field immensely challenging, albeit bewildering, to the aspiring and practicing engineers.

The book focuses on the conceptual aspects of protection and is targeted to undergraduate and graduate students of electrical engineering and researchers. Field engineers will also find the book interesting, since required knowledge of mathematics is kept to a minimum.

I owe my gratitude to Professor S. R. Bhide, a colleague of mine, who edited most of the text and prepared all the artwork.

Y. G. Paithankar

Contents

Series Introduction   *iii*
Preface   *v*

1. Basic Philosophy of Relaying   1
   - 1.1 Causes of Faults   1
   - 1.2 Types of Faults   1
   - 1.3 Effects of Faults   3
   - 1.4 Fault Statistics   3
   - 1.5 Purpose and Requirements of Protective Relays   4
   - 1.6 Relay and Circuit Breaker Locations   4
   - 1.7 A Typical Relay and Its Protective Zone   6
   - 1.8 Primary Protection, Backup Protection, and Selectivity   10
   - 1.9 Overlapping of Adjacent Protective Zones   15
   - 1.10 Problems and Exercises   17

2. Difference in Protection Requirements for Increasing Line Voltage   19
   - 2.1 Introduction   19
   - 2.2 Radial or Interconnector–Stability Problem   19
   - 2.3 Fault MVA and Subsequent Damage   20
   - 2.4 Automatic Circuit Breaker Reclosure   20
   - 2.5 Selectivity Between Adjacent Relaying   21
   - 2.6 Cost of Protection   22
   - 2.7 Difference in Protection Requirements   22
   - 2.8 Problems and Exercises   23

3. Line Protection: Overcurrent and Directional Relays   25
   - 3.1 Introduction   25
   - 3.2 Setting of OC Relays   34

3.3	Numerical Example	35
3.4	High-Set Instantaneous OC Relay Combined with DTOC/IDMT	38
3.5	Choice Between DTOC and IDMT (Short and Long Lines)	43
3.6	Hardware/Software for OC Relays	45
3.7	Directional Relays	50
3.8	Hardware/Software for Directional Relays	57
3.9	Protection of Three-Phase Lines	65
3.10	Standard Setting for Phase-Fault and Ground-Fault OC Relays	69
3.11	Directionally Controlled and Supervised Relays	70
3.12	Limitations of Relays	71
3.13	Problems and Exercises	72

4. Distance Protection: HV and EHV Line Protection — 75
   - 4.1 Introduction — 75
   - 4.2 Various Distance Relay Characteristics — 82
   - 4.3 Setting Distance Relays — 83
   - 4.4 Synthesizing Relay Characteristics — 87
   - 4.5 Effect of Power Swing on Distance Relays — 130
   - 4.6 Distance Scheme for Three-Phase Lines — 135
   - 4.7 Comparison of Various Relay Characteristics — 150
   - 4.8 Three-Stepped Distance Scheme — 153
   - 4.9 Limitations of Three-Stepped Distance Scheme — 158
   - 4.10 Out-of-Step Blocking and Tripping Schemes — 170
   - 4.11 Problems and Exercises — 178

5. Carrier Schemes for HV and EHV Lines — 181
   - 5.1 Introduction — 181
   - 5.2 Autoreclosure — 182
   - 5.3 Conditions of Relaying and Circuit Breaker for High-Speed Autoreclosure — 184
   - 5.4 Carrier Coupling — 185
   - 5.5 Carrier-Aided Distance Schemes — 186
   - 5.6 Unit Carrier Schemes — 191
   - 5.7 Problems and Exercises — 197

6. Current and Potential Transformers — 199
   - 6.1 Introduction — 199
   - 6.2 Polarity or Dot Marks — 199
   - 6.3 Current Transformers — 200

	6.4 Potential Transformers	210
	6.5 Problems and Exercises	214
7.	Basics of Differential Relays	217
	7.1 Introduction	217
	7.2 Simple Differential Scheme	217
	7.3 Biased or Percentage Differential Relay	228
	7.4 Variable-Slope or Piecewise Linear Differential Relay	230
	7.5 Problems and Exercises	231
8.	Generator Protection	233
	8.1 Introduction	233
	8.2 Types of Faults and Their Detection	233
	8.3 Problems and Exercises	244
9.	Transformer Protection	247
	9.1 Introduction	247
	9.2 Transformer Connections and Phase Shift Between Input and Output Currents	247
	9.3 Differential Protection	249
	9.4 Restricted Earth-Fault Relay	257
	9.5 Magnetizing Inrush Current	259
	9.6 Overfluxing Protection	264
	9.7 Buchholz Relay/Sudden Pressure Relay for Incipient Faults	265
	9.8 Generator/Tranformer Unit Protection	268
	9.9 Problems and Exercises	268
10.	Bus Bar Protection	269
	10.1 Introduction	269
	10.2 High-Impedance Differential Scheme	273
	10.3 Stability Ratio	278
	10.4 Check Feature and Supervisory Circuit	279
	10.5 Numerical Example	280
	10.6 Problems and Exercises	286
11.	Test Procedures, Benches, and Maintenance Schedules	289
	11.1 Introduction	289
	11.2 Static Testing	289
	11.3 Dynamic Testing	299
	11.4 Computer-Based Testing	302

11.5	Maintenance Schedule	304
11.6	Problems and Exercises	305

12. Recent Advances and Futuristic View 307
- 12.1 Introduction 307
- 12.2 Relays Based on Traveling Waves 308
- 12.3 Relays Based on Statistical Nature of Noise 313
- 12.4 Relays Augmented By Derivatives (Dynamic Relays) 313
- 12.5 Swiveling Distance Relays (Adaptive Relays) 319
- 12.6 Software for Relay Setting 319
- 12.7 Intertripping Without a Carrier 320
- 12.8 Futuristic View 321
- 12.9 Limits to Relaying 325

13. Central Computer Control and Protection 327
- 13.1 Introduction 327
- 13.2 A/D Conversion of Analog Inputs 335
- 13.3 Computer Protection 344
- 13.4 Logical Structures for Digital Protection 354
- 13.5 Design of Digital Protection and Control Devices 358
- 13.6 Adaptive Protection Systems 365
- 13.7 Expert Systems 365
- 13.8 Problems and Exercises 367

References 369

Index *379*

Transmission Network Protection

1
Basic Philosophy of Relaying

1.1. CAUSES OF FAULTS

Every power system element is subjected to a fault or a short circuit. Power system elements are generators, step-up transformers, busbars, transmission lines, step-down transformers and then distribution network feeding to loads.

The causes of faults are any of the following:

1. Healthy insulation in the equipment subjected to either transient overvoltages of small time duration due to switching and lightning strokes, direct or indirect. This causes the failure of insulation, resulting in fault current or short-circuit current. The fault current magnitude may be anywhere from 10 to 30 times the full-load or rated current of the equipment. For example, if a turboalternator has 1.0 p.u. internal voltage and the d-axis reactance is 0.1 p.u., the short-circuit current, for a three-phase fault, would be $E/Xd = 1.0/0.1$, or 10 times the rated current. Fault currents for other faults, such as L-----L, L-----L-----G and L-----G, must be calculated from symmetrical component theory. Note that the fault current magnitudes normally go down from three-phase faults to L-----L-----G faults to L-----L faults, and finally, to L-----G faults.

2. Another cause of faults is insulation aging, which may cause breakdown even at normal power frequency voltage.

3. The third cause of faults is an external object, such as a tree branch, bird, kite string, rodent, etc., spanning either two power conductors or a power conductor and ground.

1.2. TYPES OF FAULTS

The type and nature of faults in a three-phase system are normally classified as (a) phase and ground, (b) permanent, (c) transient, and (d) semitransient.

1.2.1. Phase Faults and Ground Faults

Faults involving more than one phase with or without ground are designated as *Phase faults*. Faults involving any phase with ground are called *ground faults*. Thus, a system is subject to a total of 10 types of faults. Note that an L-----L-----G fault is classified as a phase fault rather than as a ground fault.

Phase fault	No. types	Ground fault	No. types
Three-phase	1	L-----G	3
L-----L-----G	3		
L-----L	3		

1.2.2. Permanent Faults

Permanent faults are created by puncturing or breaking insulators, breaking conductors, and objects falling on the ground conductor or other phase conductors. These faults are detected by relays and trip the circuit breaker, which remains locked out.

1.2.3. Transient Faults

Transient faults are of short duration and are created by transient overvoltages. These types of faults will be dealt with in Chapter 2. Basically, this fault is caused by a flashover across the insulation due to abnormal transient overvoltages, which of course are bypassed, but results in subsequent power follow current of power frequency. The relaying system sees this fault and clears it by tripping the circuit breaker (CB). After a time the fault path is deionized, and the CB can be closed automatically to restore the supply to the equipment.

1.2.4. Semitransient Faults

Semitransient faults are created by an external object such as a tree branch or rodent. In medium-voltage lines a multishot automatic CB reclosure can burn out the object, causing a fault, restoring the equipment and improving supply reliability. Such multishot reclosures are employed only in medium-voltage lines, since the fault levels are low. They cannot be applied to high-voltage lines due to abnormal fault currents and the subsequent damage.

1.3. EFFECTS OF FAULTS

The three-phase fault is the most dangerous since it causes maximum abnormal short-circuit current. In general, if faults are not cleared rapidly, the following statements are true.

1. Generators, transformers, busbars, and other equipment are likely to be damaged due to overheating and the sudden mechanical forces developed.
2. Arcing faults invariably are a fire hazard and permanently damage the equipment. The fire can also spread in the substation unless the fault current is eliminated by suitable relaying equipment and circuit breakers. Note that the relay and CB must work together for short-circuit protection.
3. Faults can reduce the voltage profile on the entire electrical system, thereby affecting the loads. A frequency drop may lead to instability among interconnected, synchronously running generators, which, unless halted by suitable means, result in cascade tripping of generators. Hence, the purpose of interconnecting power stations (power transfer over tie lines) is lost.
4. Unsymmetrical faults result in voltage imbalance and negative sequence currents, which lead to overheating.

1.4. FAULT STATISTICS

As pointed out, a three-phase system is subjected to transient or permanent faults. The majority of L-----G faults are transient or arcing faults. One can overcome these faults by *single-shot* and *high-speed autoreclosing* to ensure supply reliability. On permanent faults the automatic CB reclosure will be unsuccessful and the CB will remain in the open position. A considerable amount of dislocation to the load takes place.

Therefore, the choice of autoreclosing depends on the statistical nature of the faults. If most faults are of a transient nature, autoreclosing is a must and will be successful. No relaying scheme can, by itself, detect whether a fault is transient or permanent. Statistically, about 80% of faults are transient and 20% are permanent. Thus, CB reclosures will always be applied, irrespective of the fault. It will be successful on transient faults, ensuring reliability, whereas it will be unsuccessful on permanent faults, leading to partial loss of supply.

1.5. PURPOSE AND REQUIREMENTS OF PROTECTIVE RELAYS

The type of failure that causes the greatest concern is the *short circuit* or *fault*. The fault may lead to various abnormal conditions on the system, such as changes in current, voltage, frequency, phase angle, rate of change of these quantities, direction of power flow (active as well as reactive), and perhaps many others, still unnoticed and unutilized to synthesize new relays. A new development is a high-impedance ground-fault relay based on fault-generated harmonics.

Most existing relays are energized by voltage and/or current supplied by current and voltage transformers. The purpose of current transformers (CTs) and potential transformers (PTs) is to reduce voltages and currents to levels manageable by the relays and to physically isolate relays from high voltage.

The basic or primary function of the relay can now be defined. A relay detects the faulty element in the integrated power system and removes it, with the help of the circuit breaker, from the remaining healthy system as quickly as possible to avoid damage and maintain security or reliability of supply in the healthy system. Note that a relay invariably goes along with the CB. If fuses are used, then fault detection and current interruption is done simultaneously.

The quality of relaying depends on its sensitivity, selectivity, speed and reliability. These will be discussed later.

1.6. RELAY AND CIRCUIT BREAKER LOCATIONS

Every power system element is fed either from one end [single-end feed (SEF)] or from both ends [double-end feed (DEF)]. For SEF conditions the CB is located at one end only, from where the short-circuit (SC) current is contributed. In case of DEF, the SC current is contributed from both ends, and thus CB are required on both ends. Let us consider some simple examples.

1.6.1. Radial Lines with Single-End-Feed

Figure 1.1 shows a single line diagram of a three-phase radial line AB fed from left-end source E_A only. There is no source at the right end of the line. This line is called a single-end-feed line since the fault current is fed from one end only. Note that there are loads on bus A and bus B, called L_A and L_B. It is quite apparent that the faulty line must be tripped from end A only,

Chapter 1

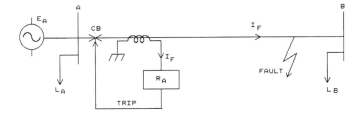

Figure 1.1 Single-end-feed line.

with the help of relay R_A, fed by fault current I_f only, whose trip output (some sort of normally open contacts) is wired to trip the CB. The relay and circuit breaker isolate the faulty line from the healthy network. Note that load L_B is lost, whereas load L_A remains undisturbed.

1.6.2. Radial Lines with Double-End-Feed

To improve the reliability of supply to L_B, it is necessary to add a source of power to end B. This is shown in Fig. 1.2. The fault current now flows from both ends and CBs are required at both ends. Both the relays and CBs must operate for the line fault to isolate the line from sources. Load L_A is fed from end A, and load L_B is fed from end B. Continuity of supply to both loads is maintained to a large extent, except that the *tie-line* power flowing from A to B, or vice versa, is lost. In fact, after the line is lost the mismatch between generation and load will be as follows:

End A $\quad P_{accelerating} = P_{tie}$

End B $\quad P_{decelerating} = P_{tie}$

Source A accelerates in speed, but the speed governor takes over and reduces its generation to L_A. The source at end B decelerates, but the speed governor increases the speed and generation (if available) to match load L_B.

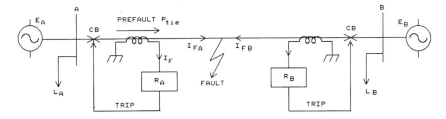

Figure 1.2 Double-end-feed line.

Basic Philosophy of Relaying

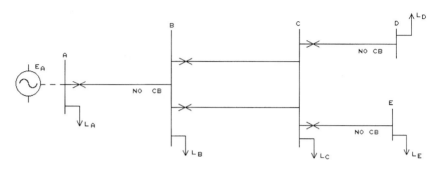

Figure 1.3 Circuit breaker locations (Example 1).

Thus, it is possible to maintain security and continuity of supply at both ends. This is the advantage of double-end-feed lines.

This example substantiates the following rule for the location of relays and circuit breakers: CBs must be at both ends of any power system element if there are sources of power at both ends. If the source of power is at one end only, then the CB must be at the source end only.

1.6.3. General Network

Now, let us take the more general example of a system (Fig. 1.3) with sources of power at some buses only. The desired CB locations are also shown. It is left to the reader to verify the correctness of the locations.

Figure 1.4 is the same as Fig. 1.3 except that CBs are placed on more busbars or ends. The correctness of these locations may be verified.

1.7. A TYPICAL RELAY AND ITS PROTECTIVE ZONE

Every relay generates its own protective zone, normally called the primary protective zone. A fault in protective zone initiates operation of the relay, and the fault is called *internal*. A fault outside the protective zone does not initiate operation of the relay, and therefore it is called an *external* or *through* fault.

1.7.1. Simple Differential Relay

Consider a simple differential relay, sometimes called a Merz Price circulating current scheme, to explain the concept of a protective zone. This relay

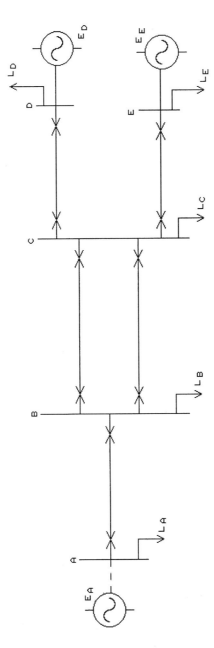

Figure 1.4 Circuit breaker locations (Example 2).

8 Basic Philosophy of Relaying

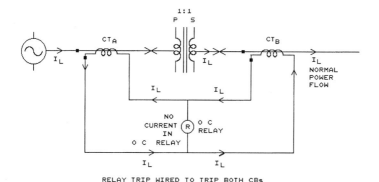

Figure 1.5 Simple differential relay.

is the basis of protecting generators, transformers and bus bars with some modifications.

Figure 1.5 shows a simple differential relay protecting a transformer fed from both ends. For simplicity, we assume that the transformer, under protection, is single phase and has a 1:1 turns ratio between the primary and secondary windings.

1.7.2. Simple Differential Relay on Load Current

It is obvious that the load currents on the primary and secondary sides will always be the same if the magnetizing current is assumed to be negligible. Under this condition of normal load flow we do not want the relay to operate.

In the figure note that there are current transformers at both ends and connected such that the CT secondary current circulates among the CTs and does not flow through the *spill* or *differential* circuit under normal power flow conditions. In the differential circuit a simple overcurrent relay (OC) is connected. Without going through the construction of the OC relay, it may be assumed that the OC relay operates (its normally open contacts will close) if there is finite current in the relay or spill or differential circuit.

Current transformers have certain dot or polarity marks. If current enters the dot mark on the primary side, secondary current exits at the dot on the secondary mark side. This ensures that the primary and secondary ampere-turns are equal and of opposite polarity, necessary for any transformer to operate (i.e., the primary and secondary ampere-turns balance).

The OC relay is wired to trip the CB open by closure of its NO contacts. For normal power flow no current flows through the OC relay; it does not operate. Thus, this differential scheme does not operate the relay

Chapter 1

and trip the CB. We do not want the transformer to be deenergized under normal power flow.

1.7.3. Simple Differential Relay on Internal Fault

In Fig. 1.6 assume a fault in the transformer. This fault is an internal fault. If we neglect the prefault load current, then the fault current I_{fA} is contributed from end A and I_{fB} is contributed from end B. In the OC relay we now have the total fault current. This operates the relay and trips the CB at end A as well at end B. Thus, the faulty power system element has been isolated from the healthy system.

1.7.4. Simple Differential Relay on External Fault

In Fig. 1.7 a fault is assumed beyond the CT at end B. This fault is not in the transformer or between the CT locations. The fault current will be the same on the primary and secondary sides, so the relay will not operate. This is correct since the fault is not in the transformer. This fault is called a through or external fault.

1.7.5. Protective Zone

The protective zone generated by the relay in Fig. 1.7 is from input CT A to output CT B, enclosing both CBs. As pointed out earlier, for an internal fault, or a fault in the protective zone, the CBs enclosed in the protective

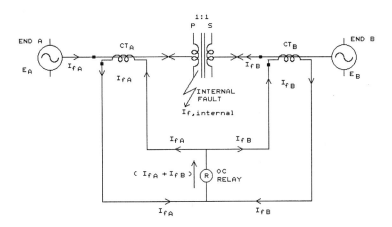

Figure 1.6 Internal fault (simple differential relay).

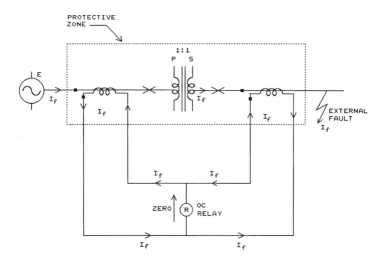

Figure 1.7 External fault (differential relay).

zone are tripped. For external or through faults or faults outside the protective zone there is no relay nor CB operation.

1.8. PRIMARY PROTECTION, BACKUP PROTECTION, AND SELECTIVITY

1.8.1. Primary Protection

Every power system element requires primary and backup protection along with CBs, the reason being the possibility of failure of the primary relaying.

Refer to Fig. 1.8. There are two radial lines, AB and BC, fed from one end only. There are local loads L_A, L_B and L_C at the three buses A, B and C. Now consider a fault on line BC fed from the source E_A. Note the correct CB locations, along with the definite time over current (DTOC) relays. A DTOC relay has input current I_r and normally open contacts as the output. The relay operates with definite time delay when I_r exceeds the pickup value. The relay has, therefore, two adjustments: (a) pickup current I_{pu}, and (b) definite operating time T_{dt}. Thus, the relay operates (normally open contact will close to trip the CB) in time T_{dt} when input current I_r exceeds the pickup value.

Relay B, along with the CB, provides primary protection to line BC. There are two possibilities to be looked into. The primary relay B and its associated CB operate satisfactorily, thereby tripping the correct CB to iso-

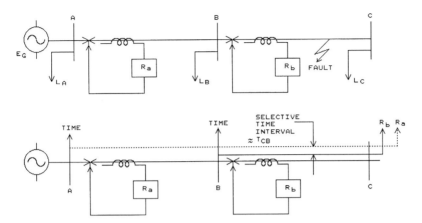

Figure 1.8 Primary protection, backup protection, and selectivity.

late the faulty line BC from the healthy system. Note that, for SEF conditions, the load L_C is lost and the other loads are unaffected. Even so, it is worthwhile to find the fault duration (in seconds) with the primary relaying working satisfactorily. If we assume that the fault on line BC takes place at $T = 0$, then the relay operates at a different time, T_r. After T_r the relay operates, initiating the opening of its associated CB. The CB begins to open, and arcing occurs in the CB male and female contacts. The fault current continues to persist until the CB interrupts the current. Let this CB time be T_{CB}. Therefore the fault duration, even if primary protection is working satisfactorily, is $T_r + T_{CB}$:

Fault duration = $T_r + T_{CB}$

1.8.2. Backup Protection

One of the many possible relay and trip circuit arrangements is shown in Fig. 1.9. It consists of current transformers, potential transformers, their lead wires to the relay, the relay itself, the trip circuit with station battery, associated leads, and the CB. If any of these components fail, the primary relaying fails. Here are causes of primary relaying failure:

1. Failure in CT and/or PT
2. Failure in the leads from CT/PT to relay
3. Failure of the relay to operate due to inadequate maintenance.
4. Failure of station battery due to inadequate charge.
5. Failure in the trip circuit.

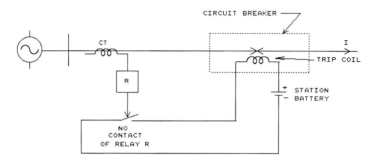

Figure 1.9 Trip circuit of circuit breaker.

6. Failure of the CB to open, perhaps due to mechanical linkage failure or welded CB contacts. The CB contacts become welded when the CB is reclosed on a fault with some bouncing effect.

1.8.3. Selectivity

Consider the possibility of failure of primary relaying at B for a fault on line BC, as shown in Fig. 1.8. The fault current persists and therefore further protection for line BC, called backup protection, is necessary. There are two types of backup protection: *local backup* and *remote backup*. Local backup is local to substation B, and the setup could duplicate the entire relaying system as primary protection. Local backup has the disadvantage that if the station battery is inadequately charged, the backup will fail to trip the CB. Dr. Mason states that "Any thing that causes the failure of primary protection should not cause the failure of backup. For local backup, the station battery, which is common to primary as well as backup could be the culprit for failure of the backup also."

Therefore remote backup is always preferred. In the single line diagram of Fig. 1.8, the correct solution to protection of line BC is to provide primary protection at B and remote backup at remote busbar A. Yet, this is not the complete solution.

Assume that the primary relaying works satisfactorily, resulting in a fault duration of $T_r + T_{CB}$ secs. If the operating time of the backup relay at station A is less than the fault duration, the backup relay would generate a trip output, tripping the CB at A also. Note that, inspite of proper operation of primary relaying, the backup also operates. This is called *loss of selectivity* or *coordination* or *discrimination* between primary and backup protection. The condition for maintaining adequate selectivity is

Chapter 1

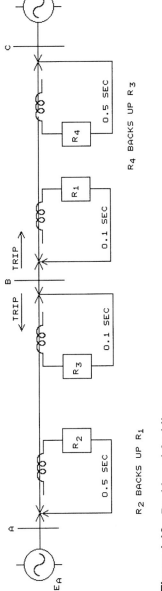

Figure 1.10 Double-end-feed line.

Basic Philosophy of Relaying

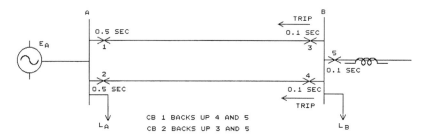

Figure 1.11 Parallel lines.

$$\text{Operating time of backup} > \text{Fault duration}$$

or

$$\text{Operating time of backup} > \text{Primary relay time} + \text{Selective time interval}$$

where selective time interval $\geqslant$ CB operating time.

Figures 1.10–1.12 show the correct locations of CBs and operating times (for selectivity) of relays for primary protection and backup protection on

1. double-end-feed lines
2. parallel lines with source at one end only
3. mesh network with source on only one bus

The trip or forward direction shown in the figures is absolutely necessary and will be dealt with while dealing with the directional relays.

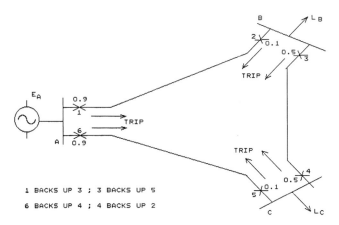

Figure 1.12 Mesh or loop network.

Chapter 1

1.9. OVERLAPPING OF ADJACENT PROTECTIVE ZONES

Figures 1.13–1.15 show the three possibilities of adjusting the protective zones for the two adjacent power system elements, in this case a line AB connected to bus A through the CB. Observe that the CB is located in the connection between the two adjacent elements. This ensures correct removal of the faulty element.

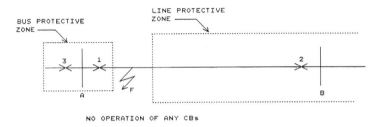

Figure 1.13 Protective zones not overlapped.

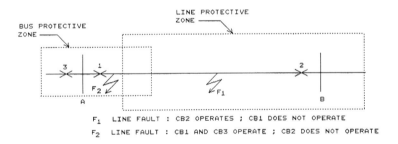

Figure 1.14 Protective zones overlapped but not around CB.

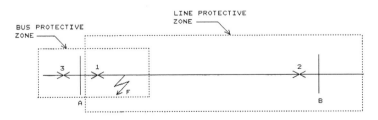

Figure 1.15 Correct overlapping of protective zones around circuit breaker.

Before we proceed, we recall the definition of protective zone: A fault in the protective zone causes the tripping of CBs located within the protective zone. In Fig. 1.13 the adjacent protective zones do not overlap. The disadvantage of this arrangement is that a fault in the nonoverlapped zone does not operate the relay and CBs. The fault is not cleared at all. This is too drastic a situation and must be avoided.

In Fig. 1.14 the adjacent protective zones are overlapped, but not around the circuit CB. A fault on line AB at fault point F_1 will operate the line relaying but not CB_1, although CB_2 operates since the line relaying protective zone does not include CB_1. For a fault F_2 on the line, just to the left of overlapped zone, the bus relaying operates, tripping the left CB_1. Unfortunately the fault is not in the line protective zone, resulting in the persistence of fault current from (CB_2 side) the right-end source. This situation is also not desirable.

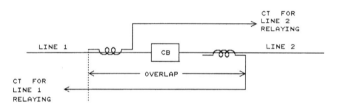

Figure 1.16 Protective zones delimited by current transformer positions.

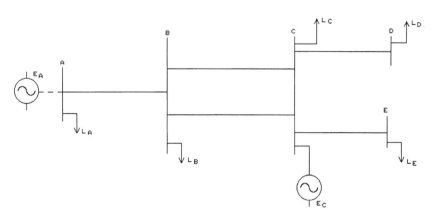

Figure 1.17 Example.

In Fig. 1.15 the protective zones are overlapped around the CB. Faults in the overlapped region trip more than the minimum number of CBs. This deenergizes both lines. This is also a drawback but is a necessary evil.

Figure 1.16 shows how the current transformers are placed around the CBs so that the protective zones overlap properly.

Some very important statements can now be made.

1. When a fault takes place in the overlapped zone more than the minimum number of CBs are tripped, causing wider dislocation in the power system.
2. There are two types of operation of backup relays.
    a. Correct operation of backup when primary fails. This also results in more than minimum CBs to trip, causing widespread dislocation to the system.
    b. Undesired operation of backup due to inadequate selectivity or discrimination or coordination between primary and backup. This is due to wrong setting of relays by the engineer, again resulting in widespread dislocation to the network.

1.10. PROBLEMS AND EXERCISES

1. Elaborate, justify, or solve the following.
    a. Circuit breakers are located between the connections to each power system element.
    b. A separate zone of protection is to be established around each power system element.
    c. Adjacent protective zones should normally overlap around the CB.
    d. A fault in a given zone will trip all CBs enclosed in the zone.
    e. If there is no overlap, a fault in the nonoverlapped region would not trip any breaker (very undesirable).
    f. A fault in the overlapped region will cause more than the minimum number of breakers to trip (wider dislocation).
    g. Why are backup relays required?
    h. State the reasons for primary relaying failure.
    i. Why is remote backup preferred to local backup?
    j. What is the fault duration when primary relaying functions satisfactorily?
    k. What is meant by selectivity/discrimination or coordination between adjacent relay locations?
    l. When backup operates, there is a wider dislocation to the power system.

2. Figure 1.17 shows an integrated network with loads and generators at various buses. Discuss the following:
    a. Draw the primary protective zones correctly around each element with suitable overlap.
    b. Identify the CBs which back up other CBs.
    c. Take faults at various locations and discuss which CBs must be tripped if (i) primary relaying functions satisfactorily, (ii) there is a failure of primary protection.

2
Difference in Protection Requirements for Increasing Line Voltage

2.1. INTRODUCTION

In an integrated power system various voltage levels are used for transmitting power from the generating stations to the end users. This is done to minimize transmission losses. The various voltage levels used are 3.3, 6.6, 11, 33, 66, 132, 220, 400, 750 kV or even higher. In this context, lines up to 66 kV are designated *low-voltage* lines, whereas lines 66 kV and above are called *high-voltage* or *extra-high-voltage* lines. This practice of designating lines varies from country to country.

A wide variety of relays are available to protect lines. They are overcurrent, distance, pilot wire differential, carrier-aided distance schemes and unit carrier schemes. The basic question is which relaying applies to which line? The problem is a natural one, since there are basic differences in protection requirements of lines as line voltage increases.

Additionally, the cost of protecting equipment has a definite relation to the cost of that equipment. For example, one does not apply expensive distance schemes to a 6.6-kV line. It is, therefore, pertinent to ask which relays should protect which line?

Before we select the type of relaying for a particular voltage transmission line, the following properties have to be kept in mind. As the line voltage increases, we can deduce the following points.

2.2. RADIAL OR INTERCONNECTOR—STABILITY PROBLEM

A low-voltage line is basically a radial line, not an interconnector between two power stations. The problem of transient stability does not exist for a radial line.

High-voltage or extra-high-voltage lines are basically tie lines or interconnectors between two power grids or power stations. Thus, there is a serious problem of maintaining transient stability between the sources. Therefore, speed of relaying is of paramount importance to HV and EHV lines, compared to LV and MV lines, for maintaining stability.

2.3. FAULT MVA AND SUBSEQUENT DAMAGE

Fault MVA and current is more for HV and EHV lines since these lines are essentially near the source of power. The LV and MV lines have much smaller fault MVA or fault current, since most of them are far away from the source. In effect, we say that the fault impedance from source to fault point is much greater in low-voltage lines compared to high-voltage lines. When we say fault MVA, we refer to a three-phase fault. Therefore, it is apparent that there is likely to be more damage to apparatuses and equipments which have to carry such abnormally high fault currents. The damage at the point of fault could be serious for faulted HV lines. Thus, it can be inferred that a high-speed relay along with a high-speed circuit breaker is a must as line voltage increases.

2.4. AUTOMATIC CIRCUIT BREAKER RECLOSURE

We have two basic types of automatic circuit breaker reclosures.

2.4.1. High-Speed, Single-Shot CB Reclosure

A high-speed, single-shot CB reclosure is absolutely necessary for high-voltage tie lines or interconnectors in order to maintain stability. Later, in Section 2.4 it will be explained that on most arcing or transient faults the auto CB reclosure is a necessity. Further, it is not desirable to have a multi-shot CB reclosure on HV lines for reasons of sudden shock and damage to the system.

2.4.2. Multi-Shot CB Reclosure in MV Lines

The purpose of a multishot CB reclosure is entirely different from that of a single-shot. It is basically meant to improve the continuity and security of supply of semitransient faults. These faults are created by an external object such as a bird, branch or rodent spanning power conductors or bus bars. These external objects can be burned by repeatedly passing fault current through them, and the fault cleared by a multishot CB reclosure. Normally, not more than two reclosures are permitted. Therefore, one must keep in mind, before selecting relaying equipment, that the purpose of highspeed and single-shot reclosure is to improve transient stability and, thus, to maintain the security of supply in HV or EHV tie lines. However, the purpose of a multishot CB reclosure in MV lines is to pass fault current repeatedly through the fault-creating object to burn it out and maintain continuity of supply.

2.5. SELECTIVITY BETWEEN ADJACENT RELAYING

Selectivity between adjacent relaying is of paramount importance as line voltage increases. Consider radial lines AB and BC between buses A-B and B-C of similar length, one a MV line and other a HV line, as shown in Fig. 2.1. Assume a load on intermediate bus B, designated as L_{MV} on MV line

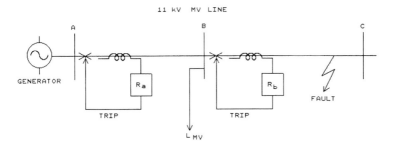

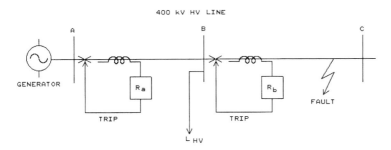

Figure 2.1 Difference in load dislocation in case of low-voltage lines and high-voltage lines.

and L_{HV} on HV line (in megawatts). Also note that, due to two different line voltages,

$$L_{HV} \gg L_{MV}$$

The power transmission capability of lines is approximately

$$P_{active} = \frac{V^2}{Z_0} \quad [MW]$$

where V = line voltage in kilovolts
 Z_0 = surge or characteristic impedance of line
      = 350–400 ohms

For AB and BC, the CB and relay positions are also shown in the figure.

In both cases presume that the relay operating time at bus A has been such that there is a loss of selectivity between relaying at A and relaying at B. The two relays are called adjacent relays. For a fault on line B-C, as shown, inspite of the fact that the primary relaying functions satisfactorily, the relay along with CB at A malfunctions due to an inadequate selective time interval. Thus, the load lost in HV lines is L_{HV} MW compared to L_{MV} MW in MV lines. Since $L_{HV} \gg L_{MV}$, there will be a wider dislocation to the loads in the HV network.

2.6. COST OF PROTECTION

More sophisticated relaying for any equipment costs more in hardware and software. The HV and EHV lines being very important lines and expensive to construct, we can afford to invest more on its protection. The percentage cost of protection is around 10% to 15% of the cost of equipment. It, therefore, appears alright, from economic point of view and the security of supply it provides, to invest more in highly sophisticated and reliable relaying for HV and EHV interconnectors.

2.7. DIFFERENCE IN PROTECTION REQUIREMENTS

The difference in line protection requirements can very lucidly be explained by Table 2.1. By studying the table we can say that as the transmission voltage increases, the speed of relaying and selectivity assumes more and more importance. As we shall see, even instantaneous overcurrent relays are inadequate for HV lines. We do get the speed of relaying (i.e., instantane-

Table 2.1 Difference in Protection Requirements

←Decreases	← Line Voltage →	Increases →
←Decreases	← Fault Damage →	Increases →
← Radial	← Radial or Tie Line →	Tie Line →
← No	← Stability Problem →	Yes →
← Not for Stability	← CB Reclosure →	For Stability →
← May Malfunction	← Backup →	Should Not Malfunction →
← Small	← Load Dislocation →	Large →
← Tolerable	← Loss of Selectivity →	Not Tolerable →

ous), but OC relays are not capable of remaining selective to variations of fault MVA and type of fault.

2.8. PROBLEMS AND EXERCISES

1. Why are backup relays necessary?
2. What is local and remote backup protection? Which is preferred and why?
3. State the reasons for failure of primary relaying.
4. Discuss the following statement: Anything that causes the failure of primary relaying should not cause the failure of backup relaying.
5. What is meant by loss of selectivity between adjacent relaying?
6. There are two types of backup operation. One is the correct operation and the other is the malfunction. Explain.
7. When backup relaying functions (right or wrong operation), more than the minimum number of CBs operate, causing wider dislocation in the power system. Discuss.
8. Backup relaying must operate with sufficient time delay to give enough time to protection, if it is capable of operating correctly.

3
Line Protection: Overcurrent and Directional Relays

3.1. INTRODUCTION

As pointed out earlier, the speed of relaying and selectivity can be sacrificed, to some extent, in LV or MV lines. This is true because the other line relays (i.e., distance, etc.) are too sophisticated and expensive.

The most apparent change that takes place for a line fault is that the postfault current is much larger than the prefault load current. Therefore, overcurrent relaying is widely used not only for lines but also for many other purposes. It is therefore pertinent to study the application of OC relays to lines, a much more inexpensive choice. There are basically three types of OC relays.

3.1.1. Definite Time OC Relays

A definite time overcurrent (DTOC) relay is a single-input relay. The line current input is fed to the current coil of the relay through the line current transformer. The output is the closure of normally open (NO) Trip contacts. The relay contacts are wired to trip the desirable CB. The Operating time versus the Input current to the relay is shown in Fig. 3.1. The connections of the relay to the line are shown in Fig. 3.2. The symbol E_g is the internal voltage of the generator, Z_s is the source impedance, which may vary due to generation changes and Z_{AB}, Z_{BC} are the line impedances of lines AB and BC, respectively. The fault currents for faults on various buses are

$$\text{Bus A fault current} = I_{fA} = \frac{E_g}{Z_s}$$

$$\text{Bus B fault current} = I_{fB} = \frac{E_g}{Z_s + Z_{AB}}$$

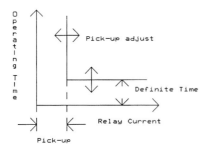

Figure 3.1 Definite time overcurrent relay.

$$\text{Bus C fault current} = I_{fC} = \frac{E_g}{Z_s + Z_{AB} + Z_{BC}}$$

The fault current versus fault position along the line is also shown in Fig. 3.2. The prefault load currents in the lines vary from time to time, and their maximum values at the two relay locations are shown as $I_{L\max,A}$ and $I_{L\max,B}$.

This DTOC relay has two adjustable settings. The first one is the Pickup value in amperes. If the current fed to the relay is less than the setting, called the pickup value, the relay does not deliver Trip output. The CB does not trip off. This pickup value is also called the plug setting. Therefore,

Pickup value = Plug setting (approximately)

Another setting for the DTOC relay is the constant or definite operating time of the relay. Note from the relay operating time versus relay input current characteristics that the relay delivers trip output only when the current exceeds the pickup value, and that after a delay, called the operating time of the relay.

a. The relay does not operate if $I_r < I_{pu}$
b. Operating time = constant if $I_r > I_{pu}$

Without bothering with the construction of the relay, we shall now consider its setting (i.e., pickup value in amperes and the definite operating time in seconds) for an example of two radial lines fed from one source only. Refer to Fig. 3.3. The single line diagram shows two lines AB and BC between buses A, B and C, with appropriate locations of DTOC relays and CBs. For simplicity, assume both CT ratios to be 1:1. Fault currents for faults on buses A, B and C are also shown. Let us now define the purpose of each relay.

Chapter 3

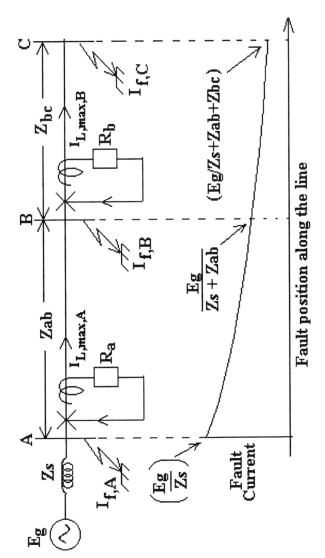

Figure 3.2 Single line diagram of two radial lines.

Line Protection

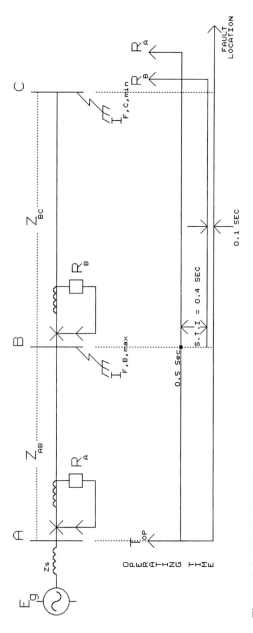

Figure 3.3 Setting DTOC relays.

Chapter 3

Relay R_B is the primary protection of line BC. A fault on line BC should be cleared fastest with the help of CB_B. Line BC is at the most remote end of the network and is not required to be selective with any other relay, since there is no other adjacent line. Therefore the following rule normally applies:

The relay at the tail end of the network is the fastest, or instantaneous. Instantaneous means without intentional time delay incorporated in the relay. Usual practice with respect to OC relaying, is instantaneous = 0.1 sec. This time takes into account the possible malfunction of the relay as a result of mechanical shock or vibration.

The operating time chosen is now 0.1 sec. Let us now choose the pickup value for R_A. The relay must operate for faults from the relay location to the end of the adjoining line (i.e., from bus B to bus C), from a backup consideration. The relay pickup value at A is now constrained by the equation

$$I_{L,\max,A} < I_{pu} < I_{f,\min,\text{bus C}}$$

The left constraint ensures that the relay does not malfunction on maximum load current at the relay location. The second constraint ensures that the relay will correctly operate for minimum fault current at the end of line BC, thus providing adequate backup protection to the entire adjacent line BC. The process is completed with the setting of I_{pu}.

Continuing with the fault on line BC, let us assume that the primary relaying for line BC is working satisfactorily. Then

Fault duration = $T_{R_B} + T_{CB}$ at B

Let relay operating time be

$$T_{R_B} = 0.1 \text{ sec (i.e., instantaneous)}$$

$$T_{CB} \text{ at B} = 0.4 \text{ sec}$$

Operating time of relay $R_A \geq 0.5$ sec

This operating time ensures the desired selectivity between R_A and R_B. Thus, the final settings for R_A are

1. $I_{L\max,A} < I_{pu} < I_{f,\text{bus C}}$
2. Operating time > 0.5 sec

3.1.2. Inverse Definite Minimum Time OC Relays

For many years, DTOC relays have been used extensively in the United Kingdom and other countries where the line lengths were small. As the line

30 **Line Protection**

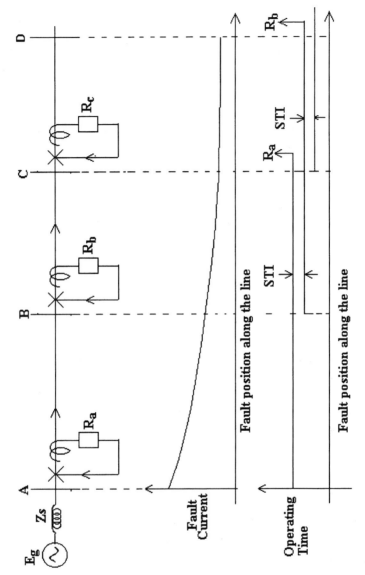

Figure 3.4 Three radial lines with DTOC relays (disadvantages).

Chapter 3

length increased, the DTOC relay was found to operate too slowly for faults close to the source, where the fault current is maximum. Before we take up the inverse definite minimum time (IDMT) relay, we look at the disadvantages of the DTOC relay.

Figure 3.4 shows three radial lines AB, BC and CD fed from end A and the settings of the three DTOC relays. The relay operating time versus fault position along the line, with proper selective interval, for relays R_A, R_B and R_C is also shown, along with the variation of fault current versus fault position. Notice that as the fault moves toward the source, the fault current increases but relay operating time also increases. This is contradictory. What we desire is that the heaviest fault must be cleared fastest.

This was the basic reason for developing the IDMT relay. It considerably improves the tripping time as the fault moves closer to the source.

The IDMT relay is single-input relay with a suitable output to trip the circuit breaker. The input-output characteristics or the operating time versus input relay current characteristics are shown in Fig. 3.5. The shaded-pole structure of this relay is shown in Fig. 3.6.

In the shaded-pole structure, a C-shaped electromagnet having two split faces is used to produce the operating torque on the aluminum disk. On one of the pole faces is a short-circuited copper band. The induced current in the copper band produces its own flux, which opposes the main flux. The resultant flux under the shaded pole has a time-phase shift with respect to the other flux under the nonshaded pole. The torque on the movable disk can be derived as follows:

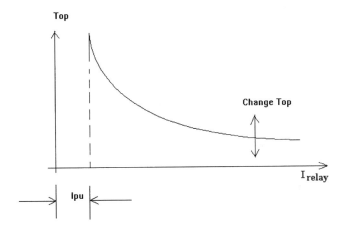

Figure 3.5 IDMT relay characteristics.

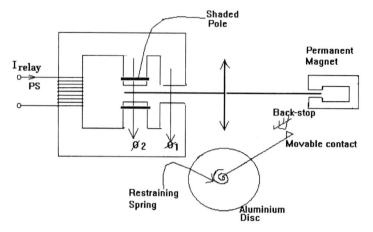

Figure 3.6 Shaded-pole structure for IDMT relay.

ϕ_1 = RMS value of flux under nonshaded pole

ϕ_2 = RMS value of flux under shaded pole

Both fluxes are directly proportional to relay current I. Hence,

Operating torque = $K_o I^2$

Figure 3.6 shows a permanent magnet, which produces a restraining torque proportional to the disk velocity. The small spiral spring ensures that the relay contacts are kept normally open.

$$T_o = T_{operating} = K_o(\phi_1 \phi_2) \sin a$$
$$T_s = T_{spring} = K_s \theta_{set}$$
$$T_d = T_{damping} = K_d \left(\frac{\theta_{set}}{T_o}\right)$$

The relay operates if

$$T_o > T_s + T_d$$

Neglecting spring torque, we get

$$\text{Operating time} = \frac{K_d}{K_o}\left(\frac{\theta_{set}}{I^2}\right)$$

where I = relay input current

K_d = damping constant due to permanent magnet

K_s = spring constant
K_o = proportionality constant for operating torque
θ_{set} = distance between fixed and movable contacts

The operating time is therefore inversely proportional to the square of the input relay current and the total distance between the fixed and movable contact.

From the torque-balance equation, the operating time is proportional to k/I^2 where k is proportional to the distance between fixed and movable contacts. Shaping the disk allows other characteristics to be generated. The conventional IDMT relay has two settings:

1. Plug setting in amperes: With this the number of turns on the current coil are selected, and therefore the pickup value of the relay. The pickup value is approximately 1.05 to 1.3 times the plug setting.
2. Time multiplier setting: With this the travel time of the disk to close the NO contacts is changed by changing the backstop of the relay, or the distance between the fixed and movable contacts, as shown in Fig. 3.6.

3.1.3. Time Multiplier Setting (TMS), Plug Setting (PS) Bridge, Plug Setting Multiplier (PSM), Pickup Errors, and Overshoot

As some of the terminology may be unfamiliar to many not closely conversant with relay techniques, the following definitions may be of value.

Time multiplier setting: A means of adjusting the movable backstop which controls the travel of the relay induction disk (for electromechanical OC relay) and thereby varies the time in which the relay will close its contacts for given values of fault current.

Plug setting bridge: It changes the turns of the current coil of the OC relay. Thus, this plug setting provides a range of pickup or plug-setting values.

Standard IDMT characteristics: Standard IDMT relay characteristics with permissible limits in time errors are shown in Fig. 3.7.

Plug-setting multiplier: It is the relay input current as multiples of plug setting. For example, if PS = 1.0 A and I_r = 20 A, then PSM = 20.0/1.0 = 20. PSM has no dimensions.

Pickup errors: Allowable error in the pickup value from the actual plug setting, as shown in Fig. 3.7.

Overshoot: The time allowed for the backup relay disk to continue to travel after the fault has been cleared by primary relaying. This value is

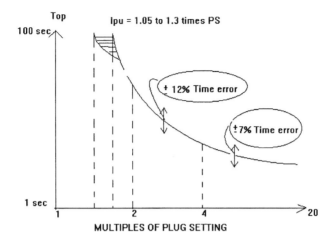

Figure 3.7 Standard IDMT characteristics.

normally taken as 10% of the actual operating time of the relay. This disk oversheet is essentially due to kinetic energy of the moving disk.

3.2. SETTING OF OC RELAYS

Figure 3.8 shows two radial lines fed from one end only. For setting the IDMT relay we consider the IDMT relay for protection of line AB and would continue with an instantaneous OC relay for line BC.

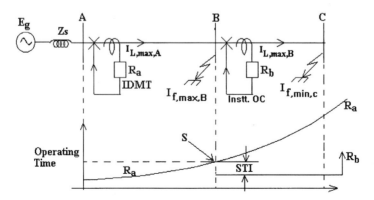

Figure 3.8 Setting OC relays.

We now have to select the pickup value and operating time (i.e., time multiplier setting) for the IDMT relay. Relay R_A has to provide backup to line BC, so its pickup value has the same constraints as the DTOC.

$$I_{L\text{max},A} < I_{\text{pu}} < I_{\text{f,min},C}$$

For relay R_A to be selective with relay R_B, the operating time of R_A for a maximum fault current on bus B should be

OT of R_A = OT of R_B + selective time interval

Selective time interval = OT of CB_B + overshoot of relay R_A

The operating time of the IDMT relay versus the fault position along the line, with selectivity being maintained with relay R_A for fault on bus B, is also shown in Fig. 3.8. If R_A is DTOC, its OT is shown by dotted lines. Note that both characteristics must pass through the same point S for maintaining selectivity. The following conclusions can now be drawn.

1. The IDMT relay provides faster primary protection to the line compared to the DTOC relay.
2. Relay R_A should be selective with relay R_B for maximum fault current for a fault on bus B. This ensures adequate selectivity.
3. Relays (IDMT or DTOC) at location A should not operate on maximum full load current at their location to avoid malfunction on line-load flow.
4. Relays (IDMT or DTOC) at location A must reach beyond bus C, for minimum fault current on bus C for adequate backup.

3.3. NUMERICAL EXAMPLE

Figure 3.9 is a single line diagram of a typical industrial network in the simplest form.

1. Line AB has a load current of 160 A with possible overload of 25%.
2. From bus B there are two outgoing lines, BC and BD. Line BC has a load current of 80 A with possible overload of 25%.
3. Line CD is protected by a fuse, whose fusing time for a fault immediately after the fuse is 0.7 sec.
4. Three-phase fault levels on various buses are

    Bus A → 3000 A   Bus B → 2000 A

5. For all CBs the operating time is 0.3 sec and overshoot may be taken as 0.1 sec.

36 **Line Protection**

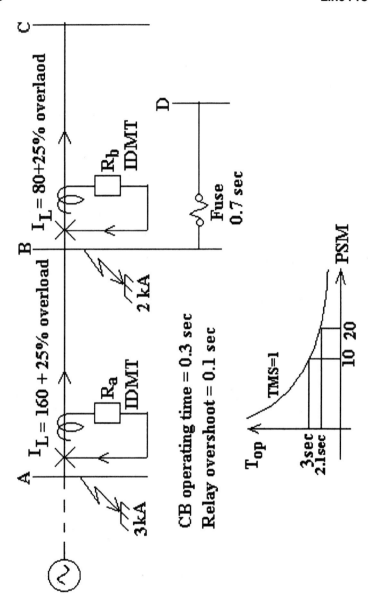

Figure 3.9 Example of setting IDMT relay.

6. The IDMT relay characteristics of a typical relay (3 sec to 10 PSM) is reproduced again for a time multiplier setting of 1.0. The TMS (which essentially adjusts the distance between fixed and movable contacts, hence the operating time) is adjustable from 0.1 to 1.0. The PS is adjusted by changing the turns on the current coil; hence, the plug settings in discrete steps are 0.5, 0.75, 1.0, 1.25, 1.5, 1.75 and 2.0 A for a 1.0-A relay.
7. Bear in mind that relay operating time is directly proportional to TMS (i.e., distance between trip contacts).
8. Assume a relay of 1.0 A. This means the current coil (CC) of the relay can carry 1.0 A continuously without damage. It is also called the thermal rating of CC. The relay CC is designed to carry about 20 times this current for a short time. This is called the short-time rating. Find the CT ratios, plug settings, and TMS for the two relays.

Solution

a. CT ratio for relay $R_B = \dfrac{80 + 0.25 \times 80}{1} = \dfrac{100}{1}$

b. CT ratio for relay $R_A = \dfrac{160 + 0.25 \times 160}{1} = \dfrac{200}{1}$

c. Since BC and BD are adjoining lines, we have to make relay R_A selective with the fuse and relay R_B for a fault current of 2000 A on bus B.

d. For relay R_B, let PS = 1.0 A (no malfunction on load) and let TMS = 0.1 sec (remote line).

e. Find the OT of relay R_B for 2000 A.

$$\begin{aligned} \text{PSM} &= \frac{\text{relay current}}{\text{PS}} \\ &= \frac{\text{fault current/CT ratio}}{\text{PS}} \\ &= \frac{2000/100}{1.0} \\ &= 20 \end{aligned}$$

From operating time versus PSM relay characteristics, the operating time for PSM = 20 and TMS = 1.0 is

Operating time = 2.1 sec (TMS = 1.0)
= 0.21 sec (since TMS = 0.1, the chosen setting of R_B

Fault duration = delay time + breaker time
= 0.21 + 0.4 = 0.61 sec

f. Relay R_A has to be selective with relay R_B as well as the fuse. The fusing time of 0.7 sec is more than the relay R_B time. Therefore relay R_A has to be made selective with the slowest protection of adjoining lines (i.e., fuse).

g. To choose the PS and TMS for relay R_A, let

PS = 1.0 (this PS avoids malfunction on loads)

Operating time of $R_A \geq$ OT of fuse (for I_f = 2000 A)

For R_A, PSM = $\dfrac{I_f}{\text{CTR}} = \dfrac{2000}{200} = 10$

The operating time of R_2 at PSM = 10 and TMS = 1 from relay characteristics is found to be 3.0 sec.

Thus, TMS of relay $R_2 = \dfrac{\text{desired OT}}{\text{TMS}} = \dfrac{0.7}{3.0} = 0.233$

Answers

For relay R_B,   PS = 1.0,   TMS = 0.10
For relay R_A,   PS = 1.0,   TMS = 0.233

3.4. HIGH-SET INSTANTANEOUS OC RELAY COMBINED WITH DTOC/IDMT

We have seen that there is a considerable reduction in tripping time if we replace DTOC with IDMT. We can further reduce this time if either DTOC or IDMT is aided by instantaneous OC relay.

The instantaneous relay, as the name signifies, operates instantaneously (i.e., without intentional time delay) as when the input current exceeds the pickup value or the plug setting. This relay, therefore, has only one setting —the pickup value in terms of plug setting in amperes. The operating time is around 0.1 sec.

Figure 3.10 shows two radial lines AB and BC fed from one end. The relays for the protection of line BC are assumed to be instantaneous, whereas for line AB they are a combination of instantaneous and DTOC for primary protection to AB and backup protection to BC. The current coils of the two relays are connected in series, whereas the trip contacts are connected in parallel, or logically or compounded, to trip the CB at A. The instantaneous OC relay is adjusted to reach 80% to 90% of line length AB, thus providing high-speed primary protection. When we say the relay is made to reach 90%,

Chapter 3

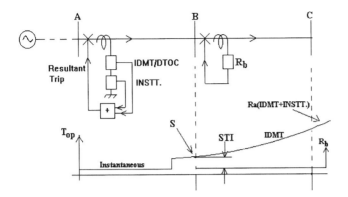

Figure 3.10 Setting instantaneous OC relay.

we mean the relay pickup value equals the fault current I_f for a fault at 90% of line AB. The DTOC/IDMT relay is adjusted in the normal fashion, as discussed earlier, to provide backup to line BC. This is shown in Fig. 3.10.

It is the instantaneous OC relay and not the DTOC/IDMT relay that provides primary protection to 80% to 90% of line AB. The remaining 10% to 20% and the entire adjoining line BC has backup protection by DTOC/IDMT. Thus, the following statements can now be made:

1. Instantaneous OC relay provides high-speed primary protection to 80% to 90% of line length. Its pickup value is larger than either DTOC or IDMT, hence, the name high-set instantaneous OC relay.
2. IDMT/DTOC provides backup protection.

The next apparent question is, why adjust the instantaneous OC relay to only 80% to 90%? Essentially the 10–20% margin is left for possible transient overreach of the instantaneous relay. If the relay overreaches, as shown in Fig. 3.11, there will be loss of selectivity on transient faults, say at point X.

3.4.1. Transient Fault

A transient fault is a fault that has offset direct current superimposed on the AC component. This dc offset depends upon the instant on the voltage waveform at which a fault takes place. This phenomenon can be explained physically rather than by solving the differential equation, as shown in Fig. 3.12.

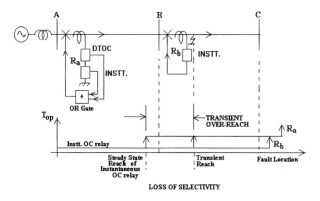

Figure 3.11 Loss of selectivity due to overreach.

The circuit shows a sinusoidal voltage source e, in series with line inductance L, resistance R and the fault switch. The differential equation is

$$e = L \frac{di}{dt} + iR$$

with initial conditions $t = 0$, $i = 0$. Solve for i.

This equation can be solved for the current i. Further, the resistance R can be neglected because it is so small. The reader can verify the following solution:

$$i = I_{max} \sin(wt + U - 90°) + I_{max} e^{-t/T} \sin(U - 90°)$$

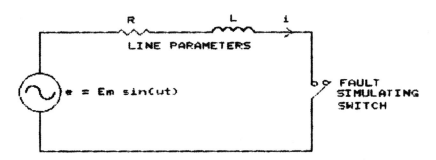

Figure 3.12 Dc offset depending on instant of fault.

where $I_{max} = E_{max}/wL$
U = voltage angle at which fault switch is closed
T = time constant = L/R = infinity if $R = 0$

(i) No dc offset: If $U = 90°$, then

$$i = \underbrace{I_{max} \sin(wt)}_{\text{ac component}} + \underbrace{\text{zero}}_{\text{no dc component}}$$

(ii) Full dc offset: If $U = 0°$, then

$$i = I_{max} \sin(wt - 90°) + I_{max} \sin(90°)$$
$$= \underbrace{-I_{max} \cos(wt)}_{\text{ac component}} - \underbrace{I_{max}}_{\text{dc component}}$$

We now explain the solution physically without resorting to mathematics. Refer to Fig. 3.13 for a fault occurring when the voltage is passing through its positive peak value. The ac sinusoidal current lags the voltage by 90° (i.e., $R \to 0$). At $t = 0$, $i = 0$, thus satisfying the initial condition. There is no dc offset nor transient in the fault current.

Now refer to Fig. 3.14, where the fault takes place when the voltage is passing through its zero value, the voltage slope being positive. The sinusoidal ac current continues to lag the voltage, but, at $t = 0$, $i = -I_{max}$, which does not satisfy the initial condition. Therefore, a positive dc offset of $+I_{max}$ is generated with decaying time constant L/R.

We now make the following statements:

1. If a fault occurs on the line at the instant the voltage waveform is passing through its positive or negative peak value, there is no transient nor dc offset in the fault current. The fault current is sinusoidal.
2. If a fault occurs when voltage is passing through its zero value, there is a dc offset superimposed on the sinusoidal fault current.

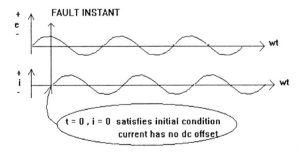

Figure 3.13 Instant of switching for no dc offset.

Line Protection

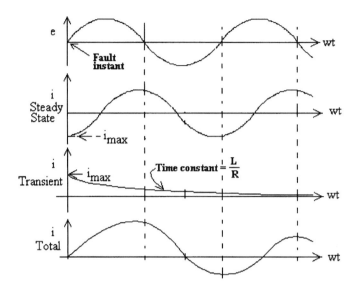

Figure 3.14 Instant of switching for dc offset.

Furthermore, if at zero voltage:
a. The voltage slope is positive, the dc offset is $+I_{max}$.
b. The voltage slope is negative, the dc offset is $-I_{max}$.

What is the effect of this decaying dc offset on relays? It is safe to assume that, if the operating time of any relay is larger than the time constant of the dc transient, the relay operation will not be affected, because the dc offset will have disappeared long before the relay operates.

High-speed relays, of course, would be affected by this dc offset. The RMS value of current seen by the relay in the first few cycles would be much more. Correctly speaking,

$$\text{RMS value} = \sqrt{I_{dc}^2 + I_{ac}^2}$$

We now consider how the high-speed instantaneous OC relay overreaches on fault currents having a dc offset.

3.4.2. Transient Overreach

Figure 3.15 shows a line emanating from bus A with a source at one end only. Let the line be protected by an instantaneous OC relay fed by line CT with a ratio of 100:1 A. A fault at point X takes place on the peak value of

Chapter 3

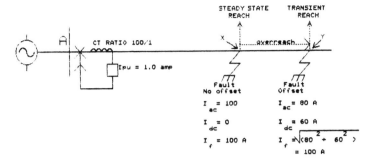

Figure 3.15 Transient overreach on offset fault current.

the voltage waveform; thus, no offset current exists in the purely sinusoidal current. Let the pickup value of the relay be 1.0 A. The OC relay, therefore, reaches point X under steady-state fault current.

Now move the fault point toward point Y, and the fault takes place at voltage zero. The fault current now has a dc offset superimposed on the ac component. The steady-state ac current will be smaller than 100 A.

Let $I_{ac} = 80.0$ A and $I_{dc} = 60.0$ A. Therefore,

$$I_{RMS} = \sqrt{6400 + 3600} = 100 \text{ A}$$

On the relay side

$$I_{RMS} = \frac{100}{\text{CT ratio}} = 1.0 \text{ A}$$

The OC relay will now operate. The relay is said to overreach from X to Y.

$$\% \text{ transient overreach} = \frac{(AY - AX)100}{AX}$$

All high-speed relays are affected by dc offset in the fault current and generally they tend to overreach. The overreach is normally taken as 10% to 20% of the actual reach. This has to be accounted for, while setting the relay and avoiding loss of selectivity with the adjacent line.

3.5. CHOICE BETWEEN DTOC AND IDMT (SHORT AND LONG LINES)

If IDMT provides faster protection, the question arises, what is the use of DTOC? There is a rule that short lines are protected by DTOC; for long lines IDMT is preferred over DTOC (Fig. 3.16).

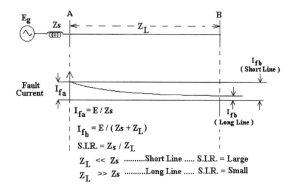

Figure 3.16 Choice between DTOC and IDMT.

In Fig. 3.16 the fault currents for a fault at bus A (i.e., the beginning of the line) and for a fault at bus B (i.e., at the end of the line) are

$$I_{f,A} = \frac{E}{Z_s} \quad \text{for fault on bus A}$$

and

$$I_{f,B} = \frac{E}{Z_s + Z_1} \quad \text{for fault at bus B}$$

Now we define the system impedance ratio (SIR) as $Y = Z_s/Z_1$. The figure shows the two fault currents against the fault position along the line. It is apparent that there will be no substantial difference in the two currents if Z_1 is small compared to Z_s. In relaying practice we define short lines and long lines by the system impedance ratio.

Therefore on short lines, where Z_s/Z_1 is large, the difference in the two fault currents is small. There is no substantial reduction of relay operating time, as the fault is moved from remote end B to near end A. The inverseness of the IDMT relay is not being utilized. Thus, for short lines the less expensive DTOC is preferred. On long lines (i.e., SIR = $Z_s/Z_1 \rightarrow$ small), the difference in the two fault currents is large and we prefer the more expensive IDMT relay. The inverseness of this relay considerably reduces the operating time as the fault is moved from the remote end to the near end of line. The practice is

For short lines (Z_s/Z_1 = SIR = large), DTOC is preferred.
For long lines (Z_s/Z_1 = SIR = small), IDMT is preferred.
Further, the practicing engineer prefers DTOC for SIR < 2.0 and IDMT for SIR > 2.0.

3.6. HARDWARE/SOFTWARE FOR OC RELAYS

The general form of the analytical relationship between operating time versus input current, depending on inverseness, is

$$T = \frac{K}{I^n - 1}$$

where K = time multiplier setting
I = input current to OC relay in terms of multiples of plug setting
n = constant deciding the inverseness

For different values of n, the characteristics are

$$T = \frac{K}{I^0} \quad (n = 0), \qquad \text{DTOC relay}$$

$$T = \frac{0.14K}{I^{0.02} - 1} \quad (n = 0.02), \qquad \text{IDMT relay}$$

$$T = \frac{13.5K}{I^{1.0} - 1} \quad (n = 1.00), \qquad \text{very inverse}$$

$$T = \frac{80.0K}{I^{2.0} - 1} \quad (n = 2.00), \qquad \text{extremely inverse}$$

3.6.1. Electromechanical Induction Disk OC Relay

The operating principle of this relay was explained in Section III and needs no repetition.

3.6.2. Static OC Relay

There are various versions and only one is dealt with in this section and the rest will be expanded in Chapter 4.

Figure 3.17 shows the static OC relay circuit, employing diodes, transistors, operational amplifiers, etc. In the figure the line current is stepped down by an auxiliary CT and then rectified and filtered by a full-wave rectifier bridge. Thus, this dc voltage V_f is directly proportional to line current. To derive the dc voltage V_{cc} for the transistor circuitry, V_f is fed to a three-terminal voltage regulator IC chip or a zener diode, which delivers constant voltage V_{cc}, irrespective of the input voltage.

With the help of a potentiometer, a portion of voltage V_f is tapped and fed to level detector 1 (in the form of a Schmitt trigger). The potentiometer alters the plug setting. The voltage V_f, which is proportional to the fault

Line Protection

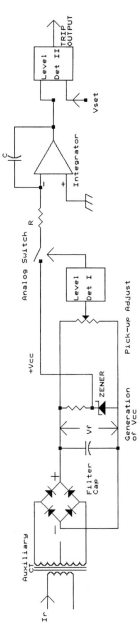

Figure 3.17 Static OC relay.

current, is fed to an op amp integrator through an analog switch, controlled by level detector 1. The integrated output from the op amp is then fed to level detector 2 and compared with voltage set value V_{set}. The output of this level detector serves the purpose of trip output. The analytical relationship between time and fault current is as follows:

1. Plug setting set by potentiometer.
2. Assuming constant current charging of integrator's capacitor C,

$$V_c = \left[\frac{1}{C}\right] I_r t > V_{set} \rightarrow \text{trip}$$

3. $T_{op} = (V_{set} C)/I_r$

The operating time versus fault current is therefore an inverse characteristic.

3.6.3. Processor-Based OC Relay

The simplest possible interface to processor circuit is shown in Fig. 3.18, and the flowchart for the implementation of the characteristic is shown in Fig. 3.19. The input line current is fed to a load R_B, across which the ac voltage is proportional to the line current. The peak value of this voltage should not exceed ±10.0 V, the bipolar A/D converter's input rating, under maximum expected input fault current. This bipolar voltage is periodically sampled at the rate of N samples per power frequency cycle. To get the RMS value of the current, these samples are squared, added, averaged, and then the square root is taken:

$$I_{RMS} = \sqrt{\frac{1}{N} \sum_{n=1}^{n=N} i_n^2}$$

The flowchart further shows that this RMS value is now compared with the pickup values read from an external keypad interfaced to an input port, and if exceeded the desired equation is solved for Time = T, as a function of TMS K and inverseness n. Alternatively the operating time T can be found from the stored relay characteristics in the form of a lookup table.

For changing inverseness and TMS, the values of n and K can be read from keypads interfaced to an input port. This time count is loaded in the timer and count down starts. If the count is zero, then trip output can be produced only if the fault current continues to be more than the pickup value. Therefore, the fault current is again checked against the set pickup value and trip issued on one of the output port pins of the processor.

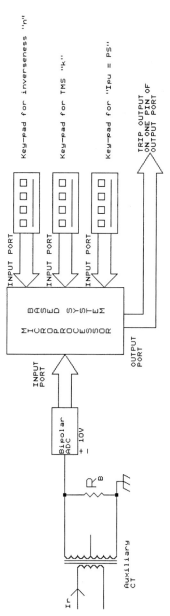

Figure 3.18 Interfacing of fault current to processor.

Chapter 3

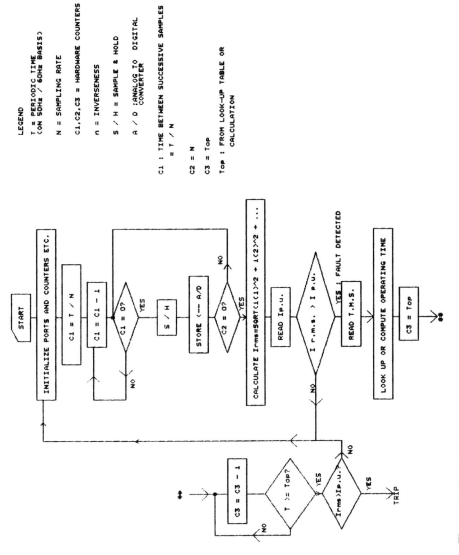

Figure 3.19 Flowchart for processor-based OC relay.

The fundamental advantage of processor-based relays is that any characteristics can be produced by choosing inverseness number n from a presettable keypad. Thus one relay can produce several OC relay characteristics.

3.7. DIRECTIONAL RELAYS

Conventional OC relays are basically nondirectional, which means the relay operates on current magnitude and not on its direction or phase shift. The relay cannot detect whether the fault is in front of or behind itself. In practice, the relay should operate in the forward direction or trip direction and the relay's operation should be blocked for a fault in the reverse or nontripping direction. The tripping direction is generated by an additional relay, called the directional relay, aiding the conventional nondirectional OC relay.

3.7.1. Where We Need Directional Relays

Figure 3.20 shows two radial lines AB and BC fed from both ends (double-end-feed lines). The circuit breakers and relay locations are also shown. Due to source of power at both ends, CBs are required at both ends of the lines. Assume all relays to be DTOC for simplicity.

We now ask

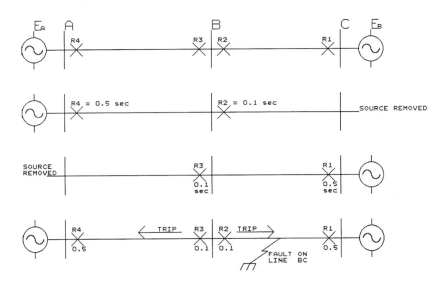

Figure 3.20 Locations of directional relays.

Chapter 3

1. Which CB backs up which CB?
2. What are the operating times of various relays, keeping in mind the selectivity?
3. Which OC relays have to be directionalized with the help of additional directional relays? Without knowing anything about directional relays (DR), we presume that DR generates a tripping direction. In other words, the directional OC relay operates only if the fault is in the tripping direction and the current to the relay exceeds the pickup value.

For setting the relay operating times the rule is to remove one source, say the right-hand source. Then the

1. OT of remotest relay $R_2 = 0.1$ sec.
2. OT of backup relay $R_4 = 0.5$ sec.

Remove the left-hand source. Then the

1. OT of remote relay $R_3 = 0.1$ sec.
2. OT of backup relay $R_1 = 0.5$ sec.

To decide which OC relays need to be directionalized, make a fault on line BC.

1. Relays R_1 and R_2 must operate to isolate the faulty line.
2. Considering the operating time of various relays, we find that the fastest relays (operating time = 0.1 sec) R_2 and R_3 would operate, thus resulting in unnecessary loss of load on bus B.
3. Therefore, the relays at R_2 and R_3 need to be directionalized as shown in the figure. Relays R_2 and R_3 operate for faults in the tripping direction.

3.7.2. Basic Principle of Directional Relays

As pointed out, the conventional overcurrent relay is incapable of detecting the direction of the fault. The basic principle of detecting fault direction can be explained by taking a single-phase case and extending it to three-phase lines.

Figure 3.21 shows two sources connected over a tie line along with the relay location. Note that the fault is in the forward or tripping direction. Note carefully the polarity marks on CT and PT and the positive direction of the relay current and relay voltage fed to the directional relay. We now have to determine how the directional relay detects the fault direction. Figure 3.22 is similar to Fig. 3.21, except that the fault is in the reverse, or nontripping, direction.

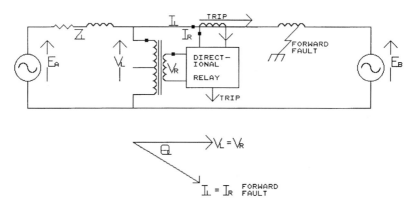

Figure 3.21 Fault in forward direction.

Note that for a fault in the tripping direction the relay current I_r lags the relay voltage V_r by the impedance angle. For a fault in the nontripping direction the fault current direction is reversed (i.e., more or less goes through a phase shift of 180°) and the phasor position of I_r with respect to V_r is now changed. These are shown in Figs. 3.21 and 3.22. Therefore, to synthesize a directional relay, we must check the phase of I_r with respect to V_r and produce a trip output for a certain phase shift margin. In this simple explanation of a single-phase case, the tripping and nontripping positions of phasor I_r w.r.t. V_r are shown in Fig. 3.23 as hatched and nonhatched regions. Thus, the basic principle of a directional relay for short-circuit protection is that it should compare the phase shift between I_r and V_r and deliver trip output for a certain phase angle between the two. This is shown in Fig. 3.23.

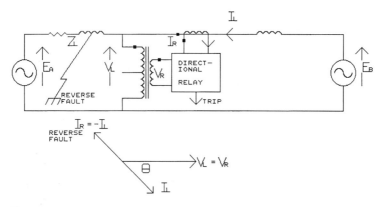

Figure 3.22 Fault in reverse direction.

Chapter 3

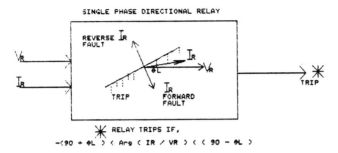

Figure 3.23 Tripping region for directional relay.

3.7.3. Phase-Fault Directional Relay for Three-Phase Lines

For three-phase lines the situation is complicated, since 3 voltages, 3 line currents and 10 types of faults (i.e., three-phase, L-L-G, L-L and L-G) are to be dealt with for faults in the tripping and nontripping direction. The question arises as to how many directional relays and what voltage and current are to be fed to which relay for phase comparison (i.e., for directional protection).

Figure 3.24 shows a diagram of a three-phase system. Assume that there is a balanced three phase power flow, at unity power factor, in the line. It is apparent that at the relay location the currents I_a, I_b and I_c will be in phase w.r.t. V_a, V_b and V_c, respectively. This is shown in Fig. 3.25.

Now for simplicity consider an A-B phase fault in the forward direction. Without resorting to symmetrical component analysis, the voltages, current and line voltage triangle will be as shown in Fig. 3.26. Roughly

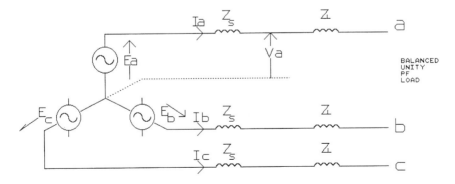

Figure 3.24 Three-phase system.

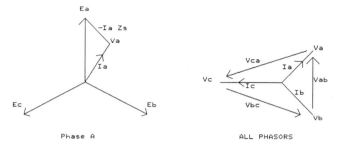

Figure 3.25 Three-phase prefault balanced situation.

speaking, the voltages and currents at the relay location for a forward fault are

$$I_a = \frac{E_{ab}}{2(Z_s + Z_1)}$$

$$I_b = -I_a$$

$$I_c = 0 \quad \text{(neglecting prefault load current)}$$

$$V_a = E_a - I_a Z_s$$

$$V_b = E_b - I_b Z_s$$

$$V_{ab} = V_a - V_b = -I_a Z_s + I_b Z_s$$

$$= -I_a + Z_s$$

$$V_c = E_c$$

The postfault voltages and currents are shown in Fig. 3.26.

For an A-B fault it will be seen that V_{AB} tends to collapse. Instead of comparing the phase of I_a w.r.t. V_a, we compare it w.r.t. other phase or line voltages that would not collapse significantly. This is called healthy phase polarization and it is the most common practice. The A-B fault directional relay has several connections. Their excitation is shown in the following.

3.7.3.1. 90°, 60°, and 30° Connections

    (i) 90° or quadrature connections

$$V_r = V_{bc}, \quad I_r = I_a$$

    (ii) 60° connection

$$V_r = V_{bc} + V_{ac}, \quad I_r = I_a$$

Chapter 3

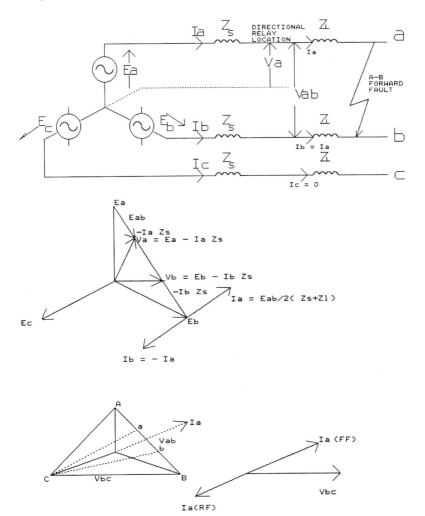

Figure 3.26 Postfault (A-B fault) voltages and currents.

(iii) 30° connection

$$V_r = V_{ac}, \qquad I_r = I_a$$

The type of connection signifies the angle between the relay voltage and current prior to a fault and at unity pf conditions.

56   **Line Protection**

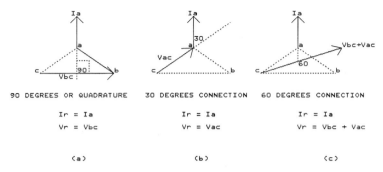

Figure 3.27   90°, 60° and 30° connections.

All 10 types of faults are taken care of by a combination of three directional relays, supervising an equal number of OC relays. The excitations to 90°, 60° and 30° A-E directional OC relay are shown in Fig. 3.27. The most common connection is the quadrature connection.

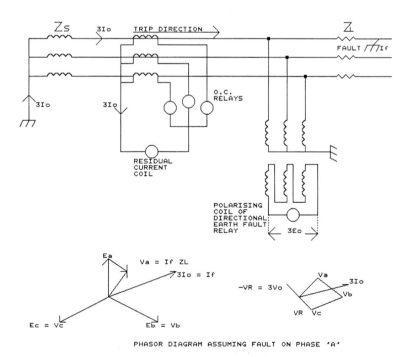

Figure 3.28   Directional relay polarization (earth fault, zero-sequence voltage).

3.7.4. Ground-Fault Directional Relays

The ground-fault directional relay detects the tripping direction of the fault. Only one relay is required and the excitations for the voltage and current coils are

V_r = zero-sequence voltage at the relay location

$$= \frac{1}{3}(V_a + V_b + V_c)$$

$$= V_0$$

I_r = zero-sequence current at the relay location

$$= \frac{1}{3}(I_a + I_b + I_c)$$

$$= I_0$$

The phase shift between zero-sequence voltage V_0 and zero-sequence current I_0 for a forward fault is shown in Fig. 3.28.

Alternatively, the directional relay can be polarized by zero-sequence current as shown in Fig. 3.29. The basic principle is that for a forward fault the two input quantities are in phase (relay trips), whereas for a reverse fault they are out of phase (relay restrains).

3.8. HARDWARE/SOFTWARE FOR DIRECTIONAL RELAYS

As stated in the previous section, the directional relay compares the phase shift between the relay voltage V_r and relay current I_r and develops trip output for a certain phase angle margin between the two for a forward fault. Such a directional relay, for short-circuit protection, is built by various techniques, such as

1. Electromechanical relay based on induction cup unit
2. Static phase comparator employing transistors or integrated circuits
3. Ring modulator using simple diodes
4. Microprocessor-based or digital-computer-based algorithm

3.8.1. Directional Relay Based on Induction Cup Unit

The basic induction cup unit is shown in Fig. 3.30. It is a four-pole induction cup structure with coils C_1 and C_2 on the two opposite pole pairs. It has an

58 Line Protection

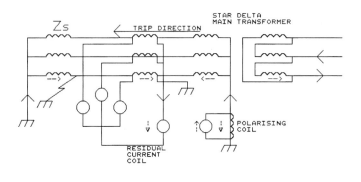

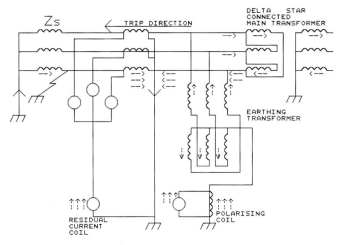

Figure 3.29 Directional relay polarization (earth fault, residual current).

inner cylindrical steel core to reduce the air gap and get a high flux density. There is a thin-walled aluminium cylinder rotating in the air gap. The travel of this induction cup is limited to a few degrees by the backstop. A spiral phosphorus bronze spring produces the restraining torque. Let these two coils produce similar frequency (i.e., 50 or 60 Hz) and sinusoidal fluxes ϕ_1 and ϕ_2 respectively (RMS values), which are in quadrature in space. Let the time phase shift between the two fluxes be a degrees. The torque on the induction cup, due to interaction of eddy currents and the flux, is

$$T_{\text{operating}} = |\phi_1| |\phi_2|(\sin a)$$

$$T_{\text{restraining}} = k \quad \text{(i.e., spring constant} \times \text{angle of travel)}$$

Chapter 3

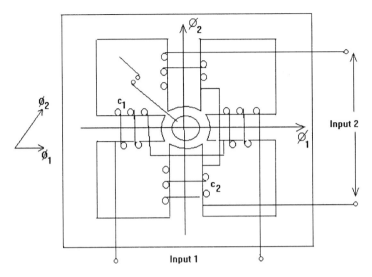

Figure 3.30 Induction cup unit.

The relay delivers trip output (i.e., movable contact touching the fixed contact) if

$$T_{\text{operating}} \geq T_{\text{restraining}}$$

This four-pole induction cup structure is a basic electromechanical unit used for developing various relay characteristics, depending on the two electrical input combinations, such as current/current, current/voltage and voltage/voltage.

For the directional relay, the inputs are relay voltage V_r and relay current I_r as shown in Fig. 3.31. Figure 3.32 shows the phasor diagram, with voltage V_r as the reference phasor.

Note that

1. Flux ϕ_v lags behind the voltage V_r by angle θ_v.
2. θ_v = impedance angle of voltage or pressure coil.
3. The flux ϕ_i is in phase with I_r.
4. θ_r = phase shift between V_r and I_r.
5. $T_{\text{operating}} = \phi_v \phi_i \sin(\theta_v + \theta_r)$.
6. Draw a line perpendicular to the pressure coil flux phasor ϕ_v. If the current phasor coincides with this line, the operating torque is a maximum, since sin(angle between ϕ_v and ϕ_i) = 1.0. Therefore this line is called the maximum torque angle (MTA) line. In other

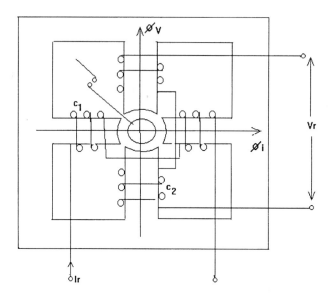

Figure 3.31 Inputs to directional relay.

words, if the phase shift between V_r and $I_r = \tau$, the maximum torque angle, the operating torque is a maximum.

7. Note that $\tau = \text{MTA} = 90° - \theta_v$.

The basic inputs to the relay structure are I_r, V_r and θ_r, the phase angle between them. The design angle of the structure is $\tau = \text{MTA} = 90° - \theta_v$.

Operating torque = $|V_r| |I_r| \cos(90° - \tau)$

The relay operates if $T_o \geq k_r$ (spring torque). The operating region or the trip region of the desired directional relay is shown in Fig. 3.33 for $k_r = 0$.

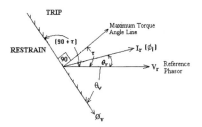

Figure 3.32 Phasor diagram of directional relay.

Chapter 3

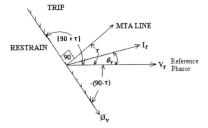

Figure 3.33 Tripping region for directional relay.

3.8.2. Directional Relay Based on Phase Comparator

A directional relay basically needs a phase-sensitive device satisfying the phase-angle criterion:

$$-(90° - \tau) < \text{Arg}\left(\frac{I_r}{V_r}\right) < +(90° + \tau)$$

where V_r = voltage at relay location
I_r = current at relay location
τ = maximum torque angle

The tripping region for the phase position of I_r w.r.t. V_r is shown in Fig. 3.34.

The phase-sensitive device, called phase comparator, using integrated circuits, is shown in a block diagram in Fig. 3.35 along with the waveform at different points.

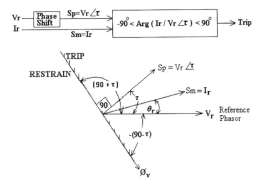

Figure 3.34 Block diagram of directional relay.

62 **Line Protection**

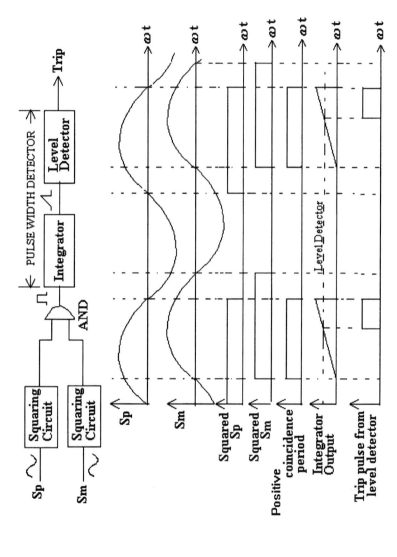

Figure 3.35 Integrated-circuit-based directional relay circuit, waveforms and phasor diagram.

Chapter 3

The detailed phase comparator is explained in Chapter 4. Briefly speaking, the relay voltage V_r is shifted through a maximum torque angle τ by a passive circuit (RLC combination) or active circuit (operational amplifier). The phase comparator inputs are then

Measuring quantity $= S_m = I_r$

Polarizing quantity $= S_p = V_r \angle{+\tau}$

These sinusoidal inputs are then waveshaped into logically squared waveforms (1 and 0) by the squaring circuit. During the positive half-cycle, it is a logical 1 and during the negative half-cycle it is a logical 0. The two squared waveforms are fed to an AND circuit, which delivers an output pulse of

Width = positive coincidence period of squared waveform

$= 180° -$ phase shift between I_r and $V_r \angle{+\tau}$

This pulse is then fed to the pulsewidth detector (i.e., a combination of integrator and Schmitt trigger), which delivers trip output if the pulsewidth $> 90°$. The waveforms at different points and the synthesized directional relay phasor diagram are shown in Fig. 3.35.

3.8.3. Directional Relay Based on Diode Ring Modulator

For simplicity we consider a half-wave diode ring modulator (Fig. 3.36) fed with

Polarizing quantity $= S_p = I_r \angle{+\tau}$

Measuring quantity $= S_m = V_r$

For a fault the current tends to increase, and the voltage tends to fall. Therefore the larger quantity I_r is chosen as the polarizing quantity and the smaller quantity $V_r \angle{\tau}$ is chosen as the measuring quantity.

During the positive half-cycle of the polarizing quantity, diodes D_1 and D_2 are in the conducting state, or on, whereas during the negative half-cycle they are nonconducting, or off. The measuring quantity S_m is, therefore, sampled during the on state and appears across the load resistance with a center tap. See Fig. 3.37.

The average value of this sampled signal of S_m during the sampling period of 0 to π degrees (1/2 cycle) is

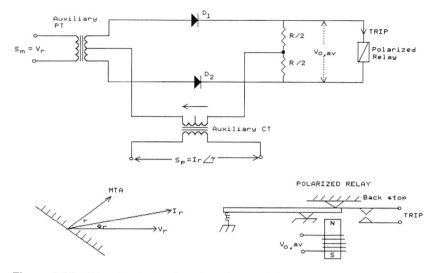

Figure 3.36 Directional relay based on ring modulator.

$$V_{av} = \int_0^\pi S_m \sin(wt - \theta_r)\, dwt$$

$$= k|V_r|\cos(\theta_r)$$

where θ_r = phase angle between S_p and S_m. The ring modulator is therefore nothing but a cosine phase comparator.

The polarized relay connected across the load resistance operates if the average value of voltage is positive and does not operate if it is negative. The polarized relay is an attracted armature-type relay, polarized by a permanent magnet. For V_{av} = positive, the flux increases and the polarized relay operates. For V_{av} = negative, the resultant flux decreases and the polarized movement remains restrained. The polarized relay is also shown in Fig. 3.36.

3.8.4. Processor-Based Directional Relay

Numerous algorithms for developing the directional relay have been reported in literature. At this juncture we draw the relay interface and the flowchart without going into details. It has been already proved, based on the phase comparator, that the cosine phase comparator, detecting the width of the positive coincidence period, can be developed as a directional relay. In other

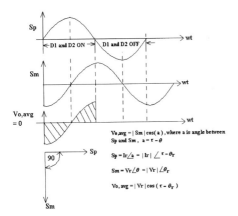

Figure 3.37 Waveforms of sampled signal S_m (ring modulator output voltage).

words, if the positive coincidence period of the two sinusoidal inputs is more than a quarter cycle (50 or 60 Hz), a cosine comparator is synthesized. Refer to Figs. 3.38 and 3.39.

The two sinusoidal inputs are fed to an A/D converter interfaced to a microprocessor. The processor detects when the voltage and current A/D ports have positive digital signals. At this instant a timer/counter in the processor will start counting up and will stop when any of the digital inputs is negative. If the time count exceeds a quarter-cycle (i.e., based on 50 or 60 Hz), a trip output logical 1 will appear on one of the data output ports.

3.9. PROTECTION OF THREE-PHASE LINES

There are two types of connections for OC protection of lines against all 10 types of faults.

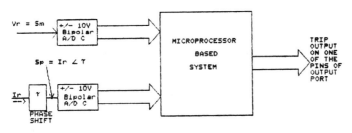

Figure 3.38 Interfacing A/D converter.

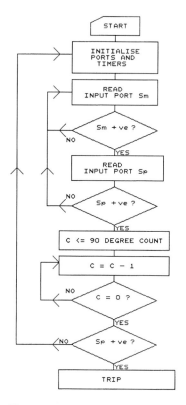

Figure 3.39 Flowchart of directional relay.

1. Three OC relays energized by line currents.
2. A combination of two OC relays energized by line currents, called *phase-fault* relays and one OC relay energized by residual current, called a *ground-fault* relay.

3.9.1. Three OC Relays Energized by Line Currents

The connections of three relays for all 10 types of faults are shown in Fig. 3.40. It is apparent that one or more relays would operate for any of the faults due to abnormal short-circuit current. The trip outputs of the relays are connected in parallel (logical OR) to trip the three-phase circuit breaker.

Chapter 3

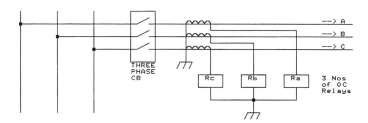

Figure 3.40 Three OC relays energized by line currents.

3.9.1.1. Data Required for Setting Three OC Relays Energized by Line Currents

The data required to set these relays (plug setting and time multiplier setting) is explained with the help of Fig. 3.41. The figure shows a single line diagram of two radial lines AB and BC protected by relays R_a and R_b respectively. For simplicity, assume R_b is instantaneous and R_a is inverse time. We choose the pickup value and operating time for R_a so that

1. Relay R_a does not function on maximum load current prior to a fault.
2. Relay R_a reaches to the end of the adjoining line BC for minimum fault current for a fault at bus C. This is necessary for adequate backup to line BC by relay R_a.
3. Relay R_a should maintain selectivity with relay R_b for maximum fault current for a fault on bus B.

These rules should be followed, since the operating time and reach of the IDMT relay depends on the source impedance (i.e., impedance behind the relay location) and type of fault (i.e., three-phase, L-L-G, L-L or L-G).

This OC scheme has serious limitations on high-impedance ground faults. Consider a high-impedance fault, such as one power conductor snapping and falling on dry ground on bus C. It may happen that

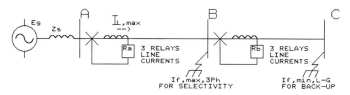

Figure 3.41 Fault data required for three OC relays energized by line currents.

68 **Line Protection**

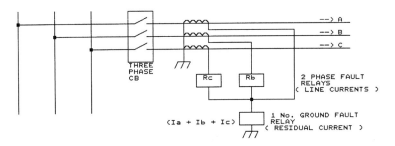

Figure 3.42 Connections of phase-fault and ground-fault OC relays.

$$I_{f,min,bus\ C,L-G} < I_{L max,bus\ A}$$

This condition violates the constraint on the pickup value of R_a, resulting in inadequate backup. The following example will make things clear. Let

$$I_{L max,bus\ A} = 100\ A \quad \text{and} \quad I_{f,min,bus\ C,L-G} = 80\ A$$

Then $100 < I_{pu}$ for $R_a < 80$, which is not possible. If $I_{pu} = 100$ A to avoid malfunction on normal load current, then the relay cannot give adequate backup to line BC for L-G faults. This limitation has led to the next OC scheme.

3.9.2. Two OC Relays Energized by Line Currents and One OC Relay Energized by Residual Current

Connections for two phase-fault OC relays energized by line currents and one ground-fault current energized by residual current are shown in Fig. 3.42.

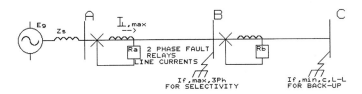

Figure 3.43 Data required for phase-fault OC relays.

3.9.2.1. Data Required for Two Phase-Fault OC Relays

See Fig. 3.43; the data required are

1. Maximum load current at the relay location $I_{Lmax,A}$ to avoid malfunction on load current.
2. Maximum three-phase fault current on bus B for selectivity of relay at bus A with relay at bus B.
3. Minimum L-L fault current at bus C, to ensure adequate backup to line BC for L-L faults.

3.9.2.2. Data Required for Ground-Fault OC Relays

Phase-fault relays are assigned the duty of clearing seven types of phase faults (three phase, L-L-G and L-L), whereas the ground-fault relay is assigned to clear three types of ground faults (L-G). As pointed out earlier, the L-G fault current at the end of the adjoining line could be less than the full load current at the relay location, thus preventing the relay from providing adequate backup. The problem can be solved by delinking the pickup value from the load current. This is done by using residual current fed to the ground-fault OC relay. If one assumes the load current to be fairly balanced on all three phases, the residual current tends to be zero. This allows us to select the pickup value of the ground-fault relay small enough to be sensitive to a L-G fault at the end of the adjoining line. The backup for L-G faults is now adequate. Thus, the data required are

1. Maximum full load current at bus A not required. Perhaps the amount of unbalanced current (i.e., residual current or zero-sequence current) is required to avoid malfunction on a small unbalance.
2. Minimum L-G fault current at bus C required for adequate backup for line B-C against L-G faults. Relay at A must reach the end of bus C for L-G faults.
3. Maximum L-G fault current at bus B for maintaining selectivity between relay at A and at B.

This is shown in Fig. 3.44.

3.10. STANDARD SETTING FOR PHASE-FAULT AND GROUND-FAULT OC RELAYS

For phase fault the plug settings for a 1-A and 5-A relay are

1. 1-A phase-fault relay: PS = 0.5, 0.75, 1.0, 1.25, and 2.0 A.

Line Protection

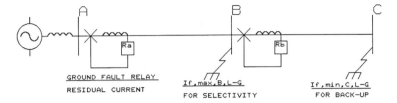

Figure 3.44 Data required for ground-fault OC relays.

2. 5-A phase-fault relay: PS = 2.5, 3.75, 5.0, 6.25, and 10.0 A.
3. Time multiplier settings for both relays: from minimum 0.1 to maximum 1.0.

3.11. DIRECTIONALLY CONTROLLED AND SUPERVISED RELAYS

There are two ways of directionalizing the simple overcurrent relay.

3.11.1. Directionally Controlled OC Relays

The OC relay has a shading coil rather than a shading short-circuited ring (Fig. 3.45). The open contacts of the shading coil are closed by the direc-

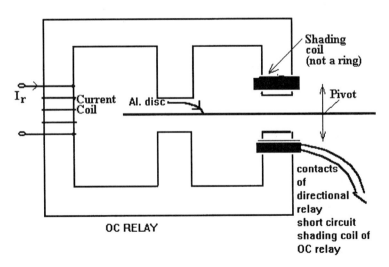

Figure 3.45 Directionally controlled OC relay.

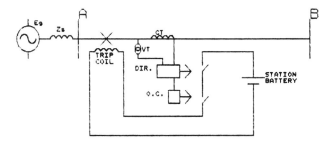

Figure 3.46 Directionally supervised OC relay.

tional relay for a forward fault, thereby starting the movement of the OC relay. For a reverse fault the directional relay contacts are open and the shading coil of the OC relay is also open-circuited. There is no phase shift between the two fluxes, and the OC relay is blocked from its operation. This type of connection is preferred worldwide.

3.11.2. Directionally Supervised OC Relays

The trip contacts of the overcurrent relay and the directional relay are connected in series (Fig. 3.46). This ensures the resultant trip only when the fault is in the forward direction and the current to the OC relay exceeds the pickup value. However, if the fault takes place in the nontripping direction and the fault current is very high, the OC relay may operate faster and before the directional relay operates. The NO contacts of the directional relay might have been closed on prefault load current. The OC relay thus may malfunction on a reverse fault also.

3.12. LIMITATIONS OF RELAYS

A single line diagram of a three-phase line with relay locations is shown in Fig. 3.47. The reach and operating time of any type of overcurrent relay depends on the fault current. We also know that for a fault at a point on the line the fault current depends on the source impedance and type of fault. Therefore, the relay reach and operating time depends on source impedance and type of fault as shown in Fig. 3.48.

Note that (a) the higher the source impedance, the lower the fault current, and the relay underreaches and the operating time increases; (b) the three-phase fault current is a maximum, whereas the L-G fault current is a minimum for a fault at the same location. Therefore, the relay overreaches and operating time decreases as the fault changes from L-G to three-phase.

Line Protection

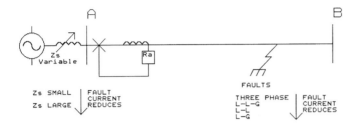

Figure 3.47 Single line diagram of a three-phase line.

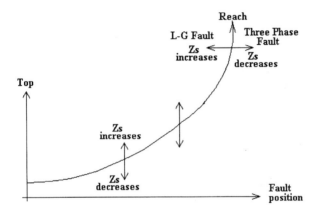

Figure 3.48 Reach and operating time of OC relays on three-phase line.

Therefore, we say that the reach and operating time of OC relays depend on source impedance and type of fault.

3.13. PROBLEMS AND EXERCISES

1. Elaborate or explain the following:
    a. The fundamental weakness of definite time overcurrent relay is that the heaviest fault (fault near the source) is cleared the slowest.
    b. ITOC relays provide faster clearing times than DTOC.
    c. When the source to line impedance ratio (Z_s/Z_L) (i.e., system impedance ratio) is large, the line is called a short line. When this ratio is small, the line is called a long line.

d. For short lines a DTOC relay is preferred, and for long lines an IDMT is preferred.
e. The speed of OC relaying can be improved by combining high-set OC along with DTOC or IDMT.
f. Why is an instantaneous OC relay called a high-set relay? Why does it tend to overreach on fault currents with dc offset or transient?
g. It is said that only high-speed relays are affected by transients, whereas slow speed relays are not affected to that extent.
h. Explain instants of fault for no offset, *positive* offset and *negative* offset in the fault current. If the fault current consists of ac and dc components, what is the approximate overreach of the instantaneous OC relay?
i. What does the time multiplier setting do in the relay?
j. What does the plug setting do in the OC relay?
   (i) What is meant by plug setting multiplier?
   (ii) What is the selective time interval?
   (iii) For two radial lines fed from one end, what is the constraint on plug setting from the point of view of
       (a) backup protection to the longest adjoining line?
       (b) avoiding malfunction on maximum load current at the relay location?
k. Consider three overcurrent relays for a three-phase line, energized by line currents, applicable to all 10 shunt faults or short circuits. What data are required to set these relays and why?
   (i) What are the limitations of the scheme for high-impedance ground faults?
   (ii) What is the remedy for these limitations?
l. Two phase-fault OC relays are energized by line currents and are assigned the duty of clearing seven types of phase faults (i.e., three phase, L-L-G and L-L). What data are required to set them.
m. One ground-fault OC relay is energized by CT secondary residual current. What data are required to set it?
n. Prove that reach and operating-time of OC relays depend on source impedance and type of fault.
o. Parallel lines, double-end-feed lines and mesh or loop networks require directional relays.
p. What are 90°, 60°, 30° connections for directional relays? Define for the phase A directional relay, the voltage and current for these connections.
q. What is the maximum torque angle line? If relay current co-

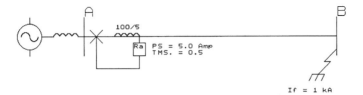

Figure 3.49 Numerical example.

incides with the MTA line, what is the torque? What is the relation between the voltage coil impedance angle and the MTA?

r. In an induction cup directional relay prove that

$$\text{Torque} = |V_r| |I_r| \cos(\tau - \theta)$$

Define τ and θ.

2. Refer to Fig. 3.49.
The 5.0-A IDMT relay has been set as follows:
CT ratio = 100/5 A, PS = 5.0 A, TMS = 0.5, fault current = 1000 A.
The operating time versus plug-setting-multiplier characteristics of the relay (3 sec to 10 PSM) at TMS = 1.0 is

PSM	2	3	4	5	10	20
Time	10	6.22	4.97	4.3	3	2.2

Find the operating time.

4
Distance Protection: HV and EHV Line Protection

4.1. INTRODUCTION

As power systems grew, bulk power needed to be transferred over long distances without great losses. The answer was to increase transmission voltage, thereby decreasing the current and consequently the losses. The power transmission capability is, approximately given by the equation:

$$\text{Active power in megawatts} = \frac{E^2}{Z_0}$$

where E = line to line voltage kV, RMS
Z_0 = surge impedance of line = 300–400 ohms

For such high-voltage lines, the overcurrent relays were found to be inadequate. This led to development of distance relays.

4.1.1. Requirements of HV/EHV Line Protection

Some protection engineers feel that OC relaying is too slow to protect high voltage lines, forgetting that we do have high-speed overcurrent relays. It is not only the speed of relaying that is important, but also the selectivity.

Chapter 2 proved that as line voltage increases the selectivity, in addition to relaying speed, assumes more importance, and this cannot be achieved by high-speed OC relays. Overcurrent relays have the inherent drawback of variable reach and variable operating time due to changes in source impedance and fault type. This limitation led to the design of a new relay whose reach and operating time are independent of the source impedance. This new relay was called the *distance relay*.

4.1.2. Principles of Distance Protection

Unlike an OC relay, which is single-input, the distance relay is two-input. The inputs are voltage and current at a point on the line, referred to as the *relaying point*. Thus, the ideal distance relay does not depend on the actual values of voltage and current, but only on their ratio and the phase angle between them.

These ideal characteristics, which define the conditions for marginal operation, are thus completely specified by the complex impedance $Z_r = V_r/I_r$. The impedance Z_r can be shown on a complex diagram having principle axes of resistance and reactance. The modulus $|Z_r|$ of the phasor Z_r, when plotted as a function of ϕ_r (the phase shift between V_r and I_r) completely defines the relay characteristics.

The locus of the impedance presented to the relay by the faulted transmission line can be superimposed on the same R-X diagram and a trip or no-trip decision taken.

Figure 4.1 shows the connections from a transmission line via current and voltage transformers to a simple impedance relay (SIR). Its characteristics are the most elementary of the various distance relays. The figure also shows the simplest form of electromechanical (EM) SIR, based on the balanced beam principle.

The EM relay operates when the operating force exceeds the restraining force. In the EM balanced beam structure, the force is proportional to ϕ^2. Therefore

Operating force = $[K_o I_r]^2$

Restraining force = $[K_r V_r]^2$

The relay trips if

$$[K_o I_r]^2 \geq [K_r V_r]^2$$

where K_o and K_r are proportionality constants. The impedance seen by the relay is

$$Z_r = \frac{V_r}{I_r} \leq \frac{K_r}{K_o}, \quad Z_r \leq \text{constant}$$

The operating characteristic for such a relay connection is also shown in Fig. 4.1 on the R-X diagram. Note that

$$R = \text{Re}(Z_r) = \text{real part} \left(\frac{V_r}{I_r}\right)$$

$$X = \text{Im}(Z_r) = \text{imaginary part} \left(\frac{V_r}{I_r}\right)$$

Chapter 4

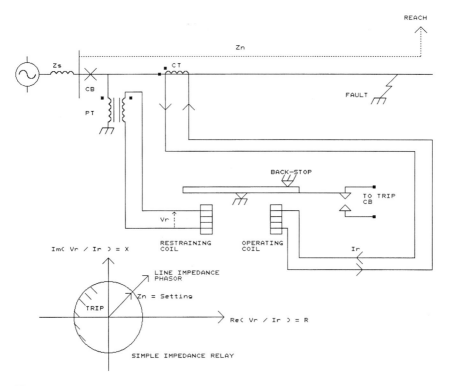

Figure 4.1 Relay connected to a line.

Consider a forward fault on the line, where $Z_n \angle \theta$ is the line impedance from the relay location up to the fault point. By Kirchhoff's law

$$V_L = I_L Z_n$$

which is independent of the impedance behind the relay location, or source impedance. Therefore,

$$\frac{V_L}{I_L} = \frac{V_r}{I_r} = Z_n \angle \theta \text{ (CT/PT ratio = 1/1)}$$

This fault impedance is now superimposed on the relay trip characteristics. The fault position is now varied so as to change $|Z_n|$. It is clear from the superimposed diagram that the relay delivers trip output for faults on the line up to line impedance = radius. The reach of the relay is also shown along the line. Such relays are called distance relays, since the fault impedance is directly proportional to the distance of the fault from the relay location.

It is imperative for us to remember that the impedance seen from the CT/PT secondary is independent of the source impedance, unlike OC relaying. Therefore, the reach of distance relays is independent of Z_s.

4.1.3. Fault Characteristics of Line on *R-X* Diagram

We define the *fault characteristic* of a line as the impedance seen from CT/PT secondary side on the faulted line. The fault position is varied from the relay location up to the desired reach point, and the faults could be metallic or arcing. The fault characteristic is plotted on an impedance or *R-X* diagram.

4.1.3.1. Single-End-Feed Line

Figure 4.2 shows an SEF line with relay location, CT, PT and variable fault position. For a metallic fault at point *F*, the impedance from CT/PT secondary side is $Z_r = Z_f$. If the fault at *F* is an arcing fault with arcing resistance R_a, then the impedance seen is $Z_r = Z_f + R_a$.

Therefore, the fault characteristic of the line will be a quadrilateral on the *R-X* diagram. It is well to remember that the impedance seen from the

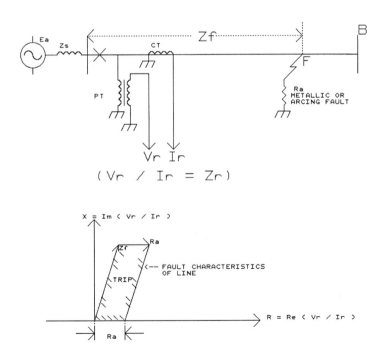

Figure 4.2 Fault characteristics of the line: single-end feed.

Chapter 4

CT/PT secondary side for both metallic and arcing faults is never outside the quadrilateral. For other than a fault condition (e.g., normal power flow, power swing, etc.), the impedance could be outside the quadrilateral.

4.1.3.2. Double-End-Feed Line with Prefault Export

Figure 4.3 shows a line AB with sources at both ends along with appropriate relay and CBs locations. Our aim is to find the impedance seen from the CT/PT secondary side for metallic and arcing faults as the fault is moved from the relay location up to the desired reach point.

It will be later proved that the relay at R_A is affected most if source A is a weaker source (Z_{sA} large) and source B is a strong source (Z_{sB} small). If the voltage E_A leads the remote end voltage E_B by an angle $+\delta$, then the power, as viewed from location A, is exported. As viewed from location B, the power is imported.

For a metallic fault at point X, the impedance is $Z_{RA} = Z_f$. For an arcing fault at X the fault current is fed into the fault from both ends. The fault

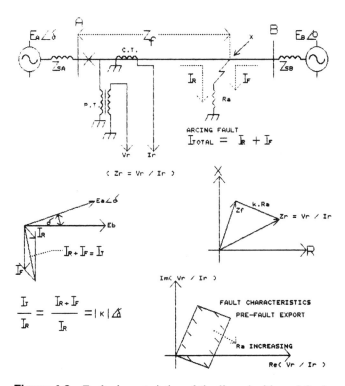

Figure 4.3 Fault characteristics of the line: double-end feed, prefault export.

80   Distance Protection

current from the remote end has much higher magnitude than the fault current fed from the local end (end A in the figure).

Assuming for brevity that all impedances are purely inductive and neglecting the arc resistance, the fault currents I_R and I_F lag the voltages E_A and E_B by 90° respectively. This is shown by the phasor diagram in the figure. Note that remote-end source current is assumed to be much greater than the local end source, because $Z_{SA} > Z_{SB}$. By applying Kirchhoff's law, we have

$$V_R = I_R Z_F + [I_R + I_F] R_a$$

Dividing by I_{RA} gives

$$\frac{V_R}{I_R} = Z_R = Z_f + \left[\frac{I_R + I_F}{I_R}\right] R_a$$

$$Z_R = Z_f + K R_a$$

and

$$K = \frac{I_R + I_f}{I_R} = \frac{\text{Total fault current}}{\text{Current on relay side}}$$

As seen from the phasor diagram, for a prefault export (i.e., angle $\delta \to$ positive), we have

$$K = |K| \angle \delta$$

where $|K| > 1.0$ and $\delta =$ positive.

As viewed from end A, the arcing resistance R_a appears to be large with a fictitious reactive component. This is shown in the figure. The fault characteristics not only enlarge but rotate clockwise w.r.t. the plain quadrilateral of the single end feed.

4.1.3.3. Double-End-Feed Line: Prefault Import

Figure 4.4 is identical to Fig. 4.3 except that δ is now negative. As viewed from relay location R_A, power is imported. The phasor diagram shows that

$$K = |K| \angle \delta$$

where $|K| > 1.0$ and $\delta =$ negative (unlike prefault export).

The fault characteristics again enlarge, but swivel counterclockwise. All three characteristics are shown in Fig. 4.5 for comparison.

Chapter 4

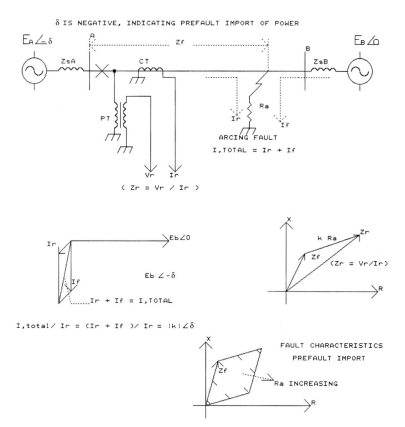

Figure 4.4 Fault characteristics of the line under double end feed, prefault import.

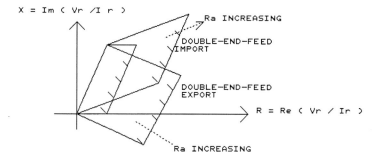

Figure 4.5 All possible line fault characteristics.

4.2. VARIOUS DISTANCE RELAY CHARACTERISTICS

4.2.1. Ideal Distance Relay: Quadrilateral

The purpose of any relay is to detect the fault and open the appropriate CB for faults only and no other conditions. Therefore, the ideal relay characteristics should be quadrilateral, so as to fit snugly around the quadrilateral fault characteristics of the line. Such characteristics are also known as *micromho* in the commercial literature, for reasons not known to the author. Figure 4.6 shows the relay characteristics. The setting Z_n and the resistive reach R along the real axis can be adjusted. If $Z_r = V_r/I_r$, the impedance as viewed from the CT/PT secondary side falls in the hatched region, and the QDR delivers trip output.

4.2.2. Simple Impedance Relay

The SIR characteristics plot as a circle on the R-X diagram with suitable radius, called the *setting*, and whose center coincides with the origin of the R-X diagram (Fig. 4.7). The adjustment available is the radius. The hatched region is the relay tripping region.

4.2.3. Reactance Relay

The reactance relay characteristics, as the name implies, have a tripping region below the setting X_n. It is a straight-line characteristic, parallel to the R-axis and offset by X_n along the X-axis. The adjustment available is the

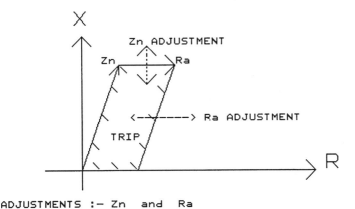

Figure 4.6 Quadrilateral distance relay characteristics.

Chapter 4

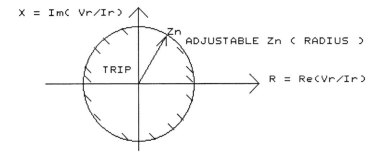

Figure 4.7 Simple impedance relay.

offset X_n. The characteristic is shown in Fig. 4.8. The hatched region is the tripping region, including the entire third and fourth quadrants.

4.2.4. MHO Relay

The MHO relay has circular characteristics, whose periphery passes through the origin on an *R-X* diagram and has diameter Z_n. The diameter has magnitude $|Z_n|$, called the setting, at angle θ_n. Available adjustments are $|Z_n|$ and θ_n. The characteristics and tripping area, which is hatched, are shown in Fig. 4.9.

4.3. SETTING DISTANCE RELAYS

The setting of distance relays, at present, means correcting the fault impedance on the line side to the relay side by current and potential transformers. Figure 4.10 shows a faulted line and CT/PT feeding into the relay.

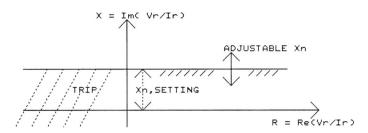

Figure 4.8 Reactance relay.

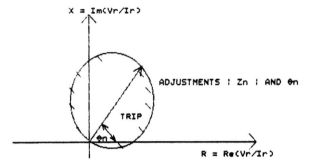

Figure 4.9 MHO relay.

4.3.1. Effect of CT/PT Ratio

Let

$$\text{Current transformer ratio} = \frac{I_L}{I_r} = \text{CTR}$$

$$\text{Voltage transformer ratio} = \frac{V_L}{V_r} = \text{PTR}$$

$$\text{Fault impedance} = \frac{V_L}{I_L} = Z_f$$

The reader can easily prove that the impedance as viewed by the relay (i.e., as viewed from CT/PT secondary site) is

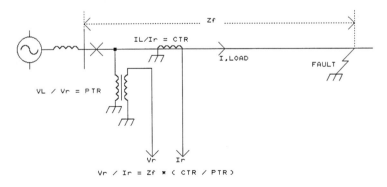

Figure 4.10 Effect of CT/PT ration on setting on distance relay.

$$Z_r = Z_f \frac{CTR}{PTR}$$

4.3.2. Quadrilateral Distance Relay

Refer to Fig. 4.11. The QDR is well defined by the points *O*, *A*, *B* and *C*. Length *OA* decides the reach of the relay along the line impedance phasor, whereas *OC* and *AB* define the accommodation of the arcing resistance for faults close to the relay location and for faults at the reach point respectively.

It is possible that arc resistance R_a may be smaller for terminal faults and larger for reach point faults due to different fault current magnitudes in the arc, as shown in Fig. 4.11. The nonlinear empirical formula for arc resistance, developed by Warrington, holds for most of the conditions and is

$$R_a = \frac{8750(S + 3ut)}{I^{0.4}}$$

where R_a = arc resistance (ohms)
  S = spacing between the conductors at which the arc is struck (feet)
  u = wind velocity (miles/hour)
  t = duration of the arc (sec)

We see that arc resistance is inversely proportional to arc current (the index being 0.4), and the arc length is $S + 3ut$, the corrected value depending upon the increase in arc length by wind velocity u over time t. The crosswind

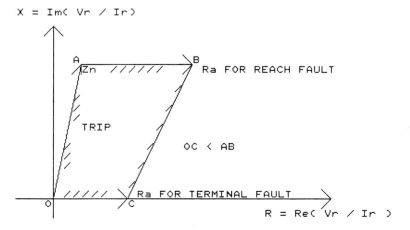

Figure 4.11 Quadrilateral distance relay.

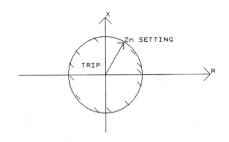

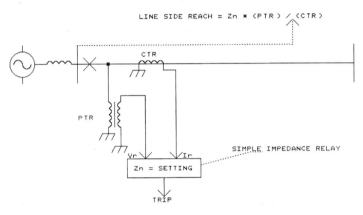

Figure 4.12 Simple impedance relay.

across the arc tends to increase its length as time increases, thereby increasing the resistance; also note that the higher the arc current the lower the arc resistance. At $t = 0$, the arc resistance equation simplifies to

$$R_a = \frac{8750S}{I^{0.4}}$$

The fault current for faults close to the relay location is more than the fault at the reach point due to increased impedance. Thus, in Fig. 4.11, $AB > OC$.

4.3.3. Simple Impedance Relay

Refer to Fig. 4.12. If the radius of SIR is Z_n, then the relay would reach to:

$$Z_{n,\text{relay side}} = Z_{f,\text{line side}} \frac{\text{CTR}}{\text{PTR}}$$

Chapter 4

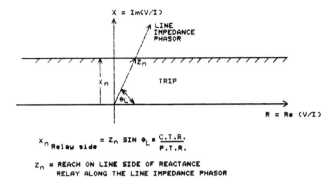

Figure 4.13 Reactance relay.

4.3.4. Reactance Relay

Refer to Fig. 4.13. The reactance relay essentially measures the reactance component of the faulted line, corrected by the CT/PT ratio. Thus, the relay setting is

$$X_{n, \text{on relay side}} = Z_n \sin(\theta_t) \frac{\text{CTR}}{\text{PTR}}$$

4.3.5. MHO Relay

Refer to Fig. 4.14, where the line impedance angle has been taken to be other than θ_n, the angle made by the diameter to the R-axis. It is quite easy to prove that

Setting of MHO = Z_n on relay side

$$= \frac{Z_{\text{reach on line side}}}{\cos(\theta_n - \theta_L)} \cdot \frac{\text{CTR}}{\text{PTR}}$$

4.4. SYNTHESIZING RELAY CHARACTERISTICS

In earlier years, all distance relays employed an induction cup unit or a balanced beam structure. The basic disadvantage of an EM relay is that it is much slower than solid-state devices or microprocessors. This is due to inertia of the moving element (i.e., induction cup or balanced beam).

The second disadvantage, which is extremely important for distance relays, is that any characteristics, other than a circle or straight line on the

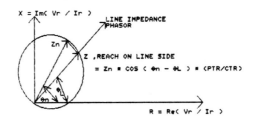

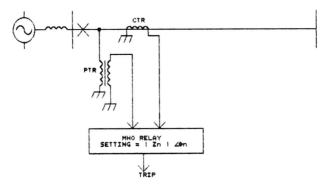

Figure 4.14 MHO relay.

R-X diagram, cannot be synthesized. Since EM relays are still being used, it is necessary to study them.

4.4.1. Electromechanical Distance Relays

4.4.1.1. Balanced Beam Structure

Figure 4.15 shows a balanced beam two-input structure. The two inputs can be voltage/voltage, voltage/current or current/current. Let us call them $S_{\text{operating}}$ and $S_{\text{restraining}}$. The structure delivers trip output (i.e., closure of normally open contacts) if

$$\frac{|S_o|}{|S_r|} > 1.0 \quad \text{(neglecting the effect of restraining spring)}$$

Chapter 4

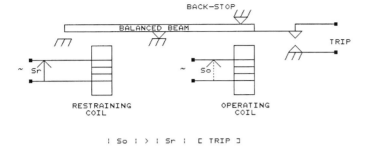

Figure 4.15 Balanced beam structure (two-input amplitude comparator).

This device is also called a *two-input amplitude comparator*, since it essentially compares the amplitude of the two input quantities, irrespective of the phase shift between them.

4.4.1.2. Induction Cup Structure

The induction cup unit is a four-pole field structure with a very thin aluminum induction cup, which carries the movable contact. This basic structure is shown in Fig. 4.16. The opposite pole pairs are fed with any combination of voltage/voltage, voltage/current or current/current. Let the RMS values of the two fluxes, which are in quadrature in space, be $|\phi_1|$ and $|\phi_2|$ with time phase shift a between them. Then the torque on the induction cup, due to interaction between flux and eddy currents, is

$$\text{Torque} = |\phi_1| |\phi_2| \sin a$$

This torque is positive for

$$0° < \text{angle } a < +180°$$

The positive torque (or tripping) and negative torque (or restraining) region is also shown in the figure. It is clear that the induction cup structure is a two-input phase comparator of the sine type.

4.4.1.3. Simple Impedance Relay

The SIR, based on a balanced beam structure, is shown in Fig. 4.17. The operating quantity is the current and the restraining quantity is the voltage, both tapped from line CT/PT respectively. The relay delivers trip output if

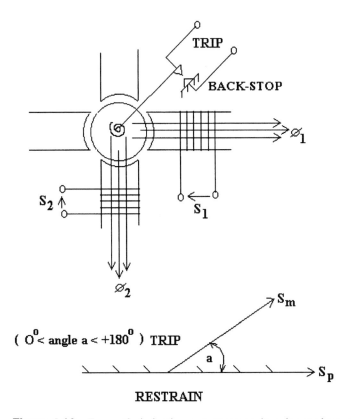

Figure 4.16 Four-pole induction cup structure (two-input phase comparator).

$$\frac{|S_o|}{|S_r|} = \frac{|K_o I|}{|K_r V|} > 1 = \frac{K_o}{K_r} > Z$$

$$\text{or} \quad Z < \frac{K_o}{K_r} \quad \text{or} \quad Z < Z_n \quad \text{where} \quad Z_n = \frac{K_o}{K_r}$$

where Z_n is the radius of the SIR, now called the *setting*, which can be varied by changing the turns on either the voltage or the current coil. The relay characteristics generated are also shown in the figure.

4.4.1.4. Reactance Relay

The reactance relay is normally built around the induction cup unit. Its characteristics are shown in Fig. 4.18. Let the impedance relay be $Z_r \angle \theta_r$,

Chapter 4

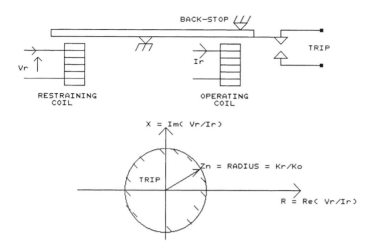

Figure 4.17 Electromechanical simple impedance relay.

where θ_r is the angle made by Z_r with the R-axis. The magnitude Z_r is now varied, and the magnitudes for trip, marginal operation and no trip are Z_{trip}, $Z_{marginal}$ and $Z_{restrain}$.

Drop the perpendiculars from the tips of the three impedances as shown. The tripping criterion is now

$$X_n > |Z_r| \cos(90° - \theta_r)$$

Letting $Z_r = V_r/I_r$ and multiplying both sides by I_r^2, we get

$$\underset{\text{operating torque}}{I_r^2 X_n} > \underset{\text{restraining torque}}{V_r I_r \cos(90° - \theta_r)}$$

The quantity I_r common to both sides is the polarizing quantity. When it is zero, the torques are also zero and the relay cannot operate. Therefore, the reactance relay is said to be *polarized* by current I_r. The current and voltage coils are placed on the four-pole field structure (Fig. 4.18) to produce the required torques. The various coils in the figure are called (a) current coil$_{\text{polarizing}}$, (b) current coil$_{\text{operating}}$ (shunted by impedance), and (c) voltage coil.

The polarizing and restraining coils, being in series, carry the same current, and therefore there is no phase shift between the two fluxes, and

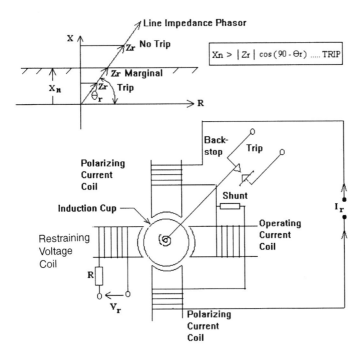

Figure 4.18 Reactance relay.

hence no operating torque. To get a finite operating torque, the restraining coil needs to be shunted by a suitable impedance so that a phase shift is developed between the polarizing and operating fluxes.

By applying the superposition theorem, let us deenergize the voltage coil (V_r not applied). Then the torque due to interaction between the polarizing flux and operating flux is

$$\text{Torque}_{\text{operating}} = k_o I^2$$

Now, if we short-circuit the operating coil, the other torque, due to interaction between polarizing and restraining fluxes, is

$$\text{Torque}_{\text{restraining}} = k_r V_r I_r \cos(90° - \theta_r)$$

where maximum torque angle = $90° - \theta_v$
θ_v = impedance angle of voltage coil
= zero (by inserting large resistance in voltage coil, see figure)

Therefore, MTA = 90° (refer to the torque relation for a directional relay) and

Torque = $VI \cos(90° - \theta_r)$

This is the desired trip criterion for a reactance relay.

4.4.1.5. MHO Relay

The MHO relay, having a setting of $OA = Z_n$ (i.e., diameter), is shown in Fig. 4.19, along with a variable magnitude of Z_r. The three values of Z_r are in the tripping region, or marginal operation, or the nontripping region. Note that

Length $OP = |Z_n| \cos(\theta_n - \theta_r)$

$OP \geq |Z_r|$ (trip)

$|Z_n| \cos(\theta_n - \theta_r) \geq |Z_r|$ (trip)

operating quantity $\geq$ restraining quantity

To implement these equations for synthesizing the MHO relay, two torques must be developed on the induction cup. Since the basic inputs at the relay location are voltage and current, the equations must be modified from impedances to voltage and current. Substituting $|Z_r| = |V|/|I|$ and multiplying both sides by $|V||I|$, we get

$$\underbrace{|V||I||Z_n|\cos(\theta_n - \theta_r)}_{\text{operating torque}} > \underbrace{|V_r^2|}_{\text{restraining torque}} \quad \text{for trip}$$

The voltage V is the common quantity to produce both torques, so the MHO relay is polarized by the relay voltage. Both torques are zero when V is zero. This happens when the fault is very close to the relay location. The remedy is to tune the polarizing circuit to 50 or 60 Hz, which ensures that the polarizing flux is maintained for a few cycles after the fault. How long it is maintained depends on the resistance of the tuned circuit.

To generate the required operating and restraining torques, the field structure is energized as shown in Fig. 4.19. The polarizing coil is a voltage coil, the operating coil is the current coil and the restraining coil is again a voltage coil. Note that there is a capacitor connected in series with the polarizing coil, which serves the following purposes:

1. It retains the polarizing flux by resonance to 50 or 60 Hz. This is also called the *memory*.
2. It creates a phase shift between the polarizing and restraining fluxes. Both coils are energized by voltage, and a phase shift must

Distance Protection

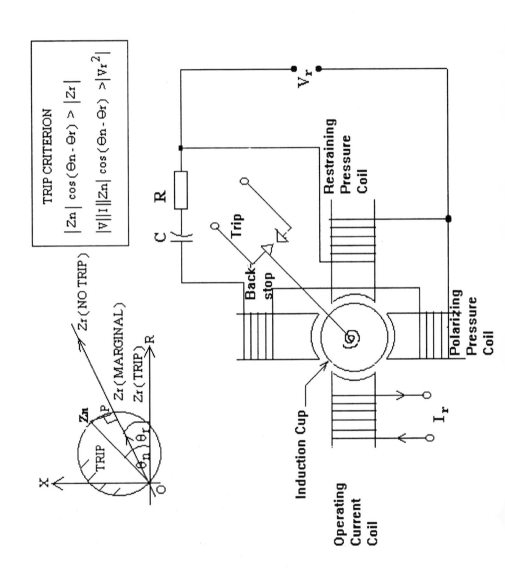

be created between the fluxes to develop a torque proportional to V^2.

Applying the superposition theorem, we first disconnect the restraining coil:

$$\text{Torque}_{\text{operating}} = |V| |I| \cos(\tau - \theta_r)$$

Adjusting the impedance angle of the polarizing circuit (i.e., θ_v) such that

$$T = \theta_n = 90° - \theta_v$$

we get

$$\text{Torque}_{\text{operating}} = k_o |V| |I| \cos(\theta_n - \theta_r)$$

Now disconnecting the pressure coil gives the restraining torque due to interaction between the polarizing and restraining fluxes:

$$\text{Torque}_{\text{restraining}} = k_r |V|^2$$

The relay operates when the operating torque is more than the restraining torque:

$$k_o |V| |I| \cos(\theta_n - \theta_r) \geq k_r |V|^2$$

Dividing both sides by $|V| |I|$ gives

$$k_o \cos(\theta_n - \theta_r) \geq k_r |Z|$$

$$|Z_n| \cos(\theta_n - \theta_r) \geq |Z| \quad \text{(tripping criterion)}$$

where $Z_n = k_o / k_r$. These are the desired MHO characteristics employing an induction cup unit.

4.4.1.6. Quadrilateral

Before the advent of comparators and microprocessors, most relays were based on a balanced beam structure or an induction cup unit, which could generate only continuous characteristics such as a circle or a straight line. Since QDR is discontinuous, its characteristics cannot be generated by electromechanical devices.

The only reference one finds is by Dr. Kimbark, who suggested a number of relays having circular characteristics sitting on top of each other, as shown in Fig. 4.20. The trip contacts were connected in parallel, resulting in characteristics somewhat loose fitting around the quadrilateral. No further attempts were made until a multi-input comparator was developed.

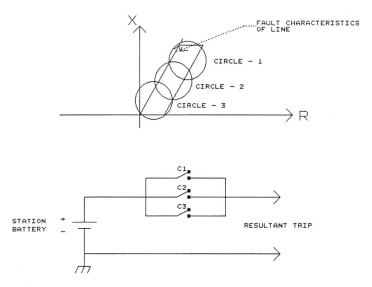

Figure 4.20 QDR as combination of circles.

4.4.2. Static Distance Relays

When uses of conventional electromechanical devices were fully explored, solid-state devices came into prominence. New protection designs were developed with a new freedom of approach that had great potential. At this time of protective gear theory, the generalized theory of phase and amplitude comparators was developed to synthesize known as well as novel desired characteristics.

Over the past few years, substantial work has been done on the use of two-input/multi-input and hybrid amplitude and phase comparators to obtain characteristics not previously attainable by induction cup units. Electromechanical relays essentially generate continuous characteristics such as a straight line or circle. Only with the help of comparators, based on solid-state devices or microprocessors, did a new range of characteristics, such as quadrilaterals and many others, become feasible.

Some of the advantages of static relays, employing diodes, transistors, thyristors, and integrated circuits are

1. Unconventional characteristics, which could not be generated by electromechanical devices.
2. High-speed operation, since there are no moving mechanical parts, having inertia.

3. Reduced volt-ampere burden by these relays on current and potential transformers. This is due to inherent amplification of input current and voltage quantities, the actual actuating power coming from the battery.
4. Improved accuracy of setting due to amplification and the devices being used, as far as possible, in the switching mode.
5. Fewer malfunctions due to mechanical shock or vibration (no moving parts).
6. Improved reliability due to solid-state devices used in the switching mode rather than the active mode. In the active mode there is problem of drift due to temperature, variation in dc supply, etc.
7. Reduction in size due to small size of devices.

The advantages of static relays outweigh some of the disadvantages such as

1. Drift in the parameters of semiconductor devices. This forces the designer to use these devices in the switching mode.
2. Possible damage to semiconductor devices due to transient voltage spikes. This problem has been solved by employing varistors, zeners, etc.
3. Possible malfunctions of static relays due to harmonics and transients in the relay inputs. To overcome this problem input filters were used. Unfortunately, the filters have energy storage components, such as capacitors, which produce secondary transients. Great care, therefore, must be taken while designing the circuit.

4.4.2.1. Two-Input Amplitude Comparators

As the name implies, the two-input amplitude comparator compares the amplitude or magnitude of the two sinusoidal relay inputs, irrespective of any phase shift between them, and delivers trip output when the RMS value of the operating quantity exceeds the RMS value of the restraining quantity. In electromechanical devices the balanced beam structure can be called the amplitude comparator.

The static version of the amplitude comparator is the well-known bridge rectifier (Fig. 4.21). Two full-wave diode bridges of reverse polarity are connected back to back. At the center a polarized relay is connected.

98 Distance Protection

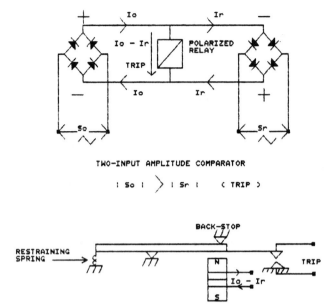

Figure 4.21 Rectifier bridge amplitude comparator.

The polarized relay is normally a very light attracted armature-type of device polarized by a permanent magnet. The polarized relay delivers trip output for a particular direction of rectified dc current. The tripping criterion is

$$\frac{|S_o|}{|S_r|} > 1.0$$

4.4.2.2. Two-Input Phase Comparator

Two-input phase comparators ideally deliver trip output for a certain phase shift between the two sinusoidal comparator inputs, irrespective of their amplitudes. In practice, the amplitudes do affect the phase angle criterion. The trip criterion, in terms of an analytical equation and graph, is shown in Fig. 4.22.

(i) **Symmetrical cosine phase comparator**

$$-90° < \text{Arg}\,\frac{S_m}{S_p} < +90°$$

Chapter 4

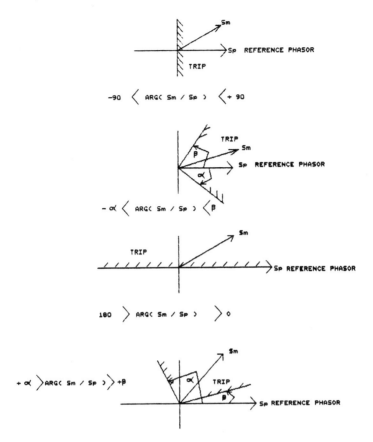

Figure 4.22 Types of phase comparators.

(ii) Asymmetrical cosine phase comparator

$$-\alpha < \text{Arg}\frac{S_m}{S_p} < +\beta$$

(iii) Symmetrical sine phase comparator

$$+180° < \text{Arg}\frac{S_m}{S_p} < 0°$$

(iv) Asymmetrical sine phase comparator

$$+\alpha < \text{Arg}\frac{S_m}{S_p} < +\beta$$

Two-input cosine comparator

The most commonly used cosine comparator is the integrating type based on coincidence principle. It can be designed with transistors or integrated circuits. Before going into detail on the circuitry, let us define the *coincidence period* (CP) of two sinusoidal inputs: The time or period during which both similar frequency sinusoidal inputs are simultaneously positive or negative. The positive and negative coincidence periods are always equal.

Figure 4.23 shows the positive and negative coincidence periods for various phase shifts. The input S_p is taken as the reference phasor, whereas the phase shift of phasor S_m is varied. Note that

1. If the two inputs are in phase, CP = half-cycle.
2. If S_m lags S_p by 90°, CP = quarter-cycle.
3. If S_m leads S_p by 90°, CP = quarter-cycle.
4. If the two inputs are out of phase, CP = zero.
5. Full cycle period = 20.000 msec for 50 Hz = 16.667 msec for 60 Hz.

The simplest circuitry employing transistors is shown in Fig. 4.24. Two transistors are connected back to back, with common collector resistance. Both transistors can be placed in the saturated state or on state by suitably adjusting the base currents. The transistor inputs are the sinusoidal quantities S_p and S_m. To avoid reverse breakdown of the base-emitter junction on the negative half-cycle of the sinusoidal inputs, diodes D_1 and D_2 are connected in the circuit. The functioning of the circuit is as follows.

1. The two transistors are in the on state during the positive half-cycle (due to forward biasing of the B-E junction); they are in the off state (due to reverse biasing of the B-E junction) during the negative half-cycle.
2. An output voltage V_{cc} appears at the common collector point A during the period when both transistors are off. The width of this output pulse equals the negative coincidence period of the two sinusoidal inputs.
3. If this pulsewidth > quarter-cycle, we have synthesized the symmetrical cosine comparator. The pulsewidth detector combines an integrator and level detector.
4. The common collector resistance R_c along with capacitor C forms a passive integrator. The maximum input pulsewidth will never be more than a half-cycle (inputs are in phase). To get relatively linear charging over the maximum period of a half-cycle, the time constant of the charging circuit R_cC is to be designed to about 10 times the half-cycle (i.e., 100 msec on a 50-Hz base). The maximum value to which capacitor C will charge is

Chapter 4

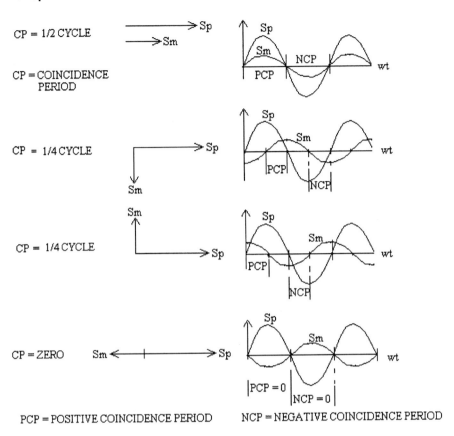

Figure 4.23 Coincidence period of two inputs.

$$V_{c,\,max} = \frac{V_{cc} \times \text{pulsewidth in msec}}{\text{Time constant}}$$

The reference level corresponding to 5.0 msec, with a time constant of 100.0 msec, is

$$V_{\text{level}\,(10\,\text{msec})} = \frac{V_{cc} \times 5.0}{100.0}$$

$$= V_{cc}/20$$

$$= 1.0 \text{ V if } V_{cc} = 20 \text{ V}$$

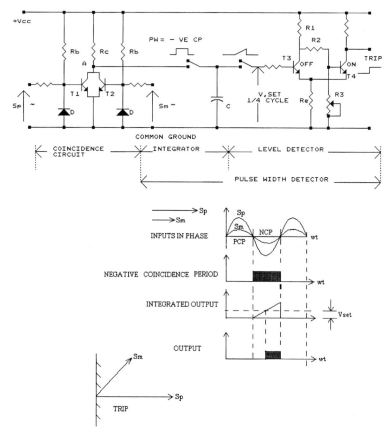

Figure 4.24 Transistorized cosine comparator based on coincidence principle.

5. To detect a voltage level of 1.0 V (corresponding to 5.0-msec pulsewidth), we use the conventional Schmitt trigger circuit as a level detector. The set level is

$$V_{\text{set for 5 msec}} = V_{cc} \times \frac{R_3}{R_1 + R_2 + R_3}$$

6. For a phase shift varying from $-90°$ to $+90°$ between the two sinusoidal comparator inputs, the pulsewidth is more than 5.0 msec, or the maximum level to which the capacitor will charge is more than $V_{\text{set for 5 msec}}$. The level detector will deliver trip output by turning off transistor T_4. The symmetrical cosine comparator has now been synthesized.

Sine comparator based on block-spike principle

Figure 4.25 shows the circuit. In this circuit, one of the sinusoidal inputs S_p is logically squared by transistor T_1 and differentiated by the C-R circuit. The voltage across resistance R will be a positive and a negative voltage spike at zero crossing of the sinusoidal inputs (i.e., slope negative and positive). The negative spike is suppressed by diode D_1. The remaining positive voltage spike is fed to the collectors of T_2 and T_3. This spike is bypassed only when the positive half-cycle of input S_m keeps the transistor pair on; else the pulse appears as output on negative half-cycle of S_m. In other words, there is a trip output in the form of a pulse if negative zero crossing of S_p is anywhere in the negative half-cycle of input S_m. The 0° to 180° sine comparator is synthesized.

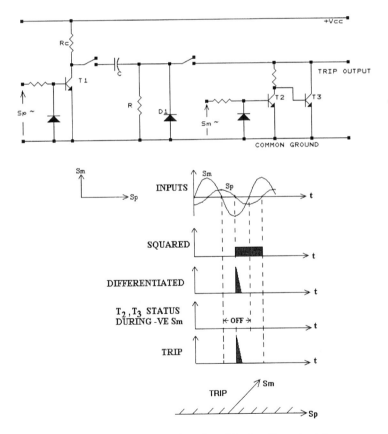

Figure 4.25 Sine comparator based on block-spike principle.

Transient performance and operating time of cosine comparator
The operating time of the cosine-type phase comparator depends on the instant the sinusoidal inputs S_p and S_m are switched on, or when these inputs go through the step change in phase shift due to a fault. Figure 4.26 shows the worst and most favorable instants of fault for the comparator to operate in maximum and minimum time, respectively. Notice that

1. Minimum operating time is quarter-cycle.
2. Maximum operating time is quarter-plus-one cycle.

To improve operating time, it is apparent that one should check for trip on the positive as well as the negative cycle. In this case the maximum operating time is a quarter-plus-half-cycle. This is called a *dual* comparator.

The transient performance of the comparator is shown in Fig. 4.27, with dc offset in one of the sinusoidal inputs. Notice that the positive coincidence period tends to increase, whereas the negative coincidence period tends to decrease from the steady-state coincidence period. Thus the comparator will operate incorrectly if only one coincidence period is checked.

The simplest solution is to get the average value of the coincidence period [i.e., (positive CP + negative CP)/2]. This period is correct if the dc offset decay is assumed negligible during one cycle. Several authors have built a dc offset-free cosine-type phase comparator. With additional

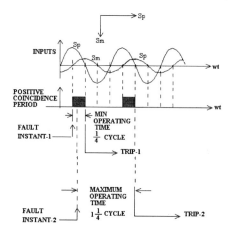

Figure 4.26 Operating times of phase comparator.

Chapter 4

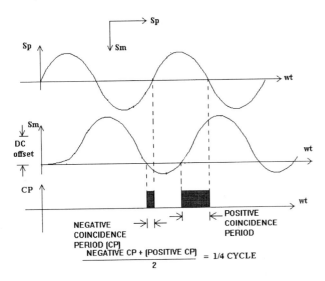

Figure 4.27 Effect of dc offset on phase comparator.

harmonics, analog filters are needed or else the entire design must use processor-based relays.

4.4.2.3. Duality Between Phase Comparator and Amplitude Comparator

A general theory, postulated by Nellist, states that any circular or straight-line characteristic synthesized by an amplitude comparator can also be synthesized by a phase comparator if the inputs of the amplitude comparator are modified to the sum and difference and fed to a phase comparator. The converse is also true. This analogy in analytical form is

$$-90° < \text{Arg} \frac{S_m}{S_p} < +90° \equiv \underbrace{|S_p + S_m|}_{\text{operating quantity}} > \underbrace{|S_p - S_m|}_{\text{restraining quantity}}$$

or

$$|S_o| > |S_r| \equiv -90° < \text{Arg} \frac{S_o + S_r}{S_o - S_r} < +90°$$

where S_o, S_r = inputs to amplitude comparator
S_p, S_m = inputs to phase comparator

To prove duality requires analytical and graphical proofs. We shall take up the graphical proof, since it is conceptually better and easier to grasp. See Fig. 4.28. There are four phasor diagrams. The phasor S_p is taken as the reference phasor.

1. In the first phasor diagram, S_m is in the first quadrant and the condition is trip for (a) phase comparator, (b) amplitude comparator, since $|S_p + S_m| > |S_p - S_m|$.
2. In the second phasor diagram, S_m is in the second quadrant and the condition is no trip for (a) phase comparator, (b) amplitude comparator, since $|S_p + S_m| < |S_p - S_m|$.
3. In the third phasor diagram, S_m is in the third quadrant and the condition is no trip for (a) phase comparator, (b) amplitude comparator, since $|S_p + S_m| < |S_p - S_m|$.
4. In the final phasor diagram, S_m is in the fourth quadrant and the condition is trip for (a) phase comparator, (b) amplitude comparator, since $|S_p + S_m| > |S_p - S_m|$.

Therefore, comparing two inputs in phase is as good as comparing their sum and difference in an amplitude comparator. The converse is also true.

4.4.2.4. Multi-Input Comparators

The generalized theory of symmetrical comparators states that no characteristics other than a circle or straight line can be generated (i.e., with any discontinuity). Thus, for any discontinuous relay characteristics we need a multi-input comparator. This is discussed in Section 4.4.2.8 where QDR is synthesized, using multi-input phase and an amplitude comparator.

4.4.2.5. Static Simple Impedance Relay

SIR based on amplitude comparator

The relay characteristics are shown in Fig. 4.29. The amplitude comparator is the commonly used rectifier bridge type with polarized relay. From the figure the inputs are

$$S_o = IZ_n, \quad S_r = V$$

The relay operates if

$$IZ_n \geq V$$

or

$$\frac{V}{I} = \text{impedance seen by the relay} = Z < Z_n$$

where Z_n = setting or SIR radius

Chapter 4

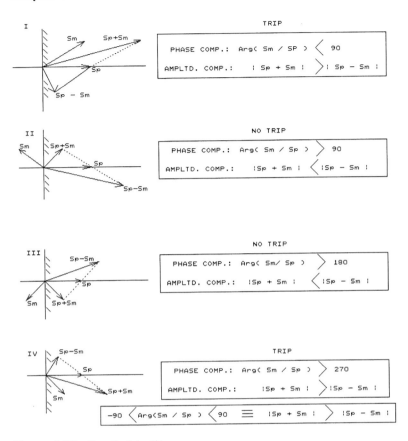

Figure 4.28 Proof of duality.

SIR based on phase comparator

Refer to Fig. 4.30, with Z in tripping, marginal and nontripping regions. In the figure, join points P_1, P_2, and P_3 with point B and point A. Note that

$$OA = BO = Z_n$$
$$BP_1 = Z_n + Z_r \quad (Z_r \text{ in trip region})$$
$$BP_2 = Z_n + Z_r \quad (Z_r \text{ for marginal operation})$$
$$BP_3 = Z_n + Z_r \quad (Z_r \text{ in nontrip region})$$
$$P_1A = Z_n - Z_r \quad (Z_r \text{ in trip region})$$
$$P_2A = Z_n - Z_r \quad (Z_r \text{ for marginal operation})$$
$$P_3A = Z_n - Z_r \quad (Z_r \text{ in nontrip region})$$

108   **Distance Protection**

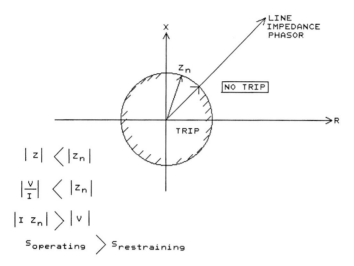

Figure 4.29 SIR based on amplitude comparator.

Note that the angle of lead of phasor $Z_n - Z$ w.r.t. $Z_n + Z$ for tripping is less than 90°. For phasor Z in the left half-circle, this angle is more than 270° or −90°. The tripping criterion, therefore, is

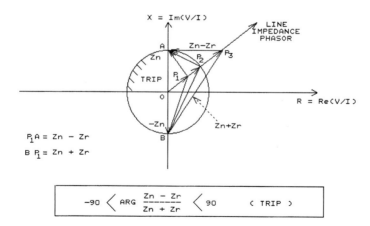

Figure 4.30 SIR based on phase comparator.

Chapter 4

$$-90° < \frac{Z_n - Z_r}{Z_n + Z_r} < +90° \quad \text{(trip)}$$

Substituting $Z = V/I$ and multiplying numerator and denominator by I, we get

$$-90° < \frac{IZ_n - V}{IZ_n + V} < +90° \quad \text{(trip)}$$

The cosine-type phase comparator described above can be used.

4.4.2.6. Static Reactance Relay

Reactance relay based on phase comparator
Refer to Fig. 4.31, which shows the reactance characteristics with $Z = V/I$ in tripping, marginal and nontripping regions. Join point A with P_1, P_2 and P_3. Then

$$P_1A = X_n - Z \quad (Z \text{ in trip region})$$
$$P_2A = X_n - Z \quad (Z \text{ for marginal operation})$$
$$P_3A = X_n - Z \quad (Z \text{ in nontrip region})$$

The angle of lead of P_1A and P_2A w.r.t. phasor X_n is less than 90° for tripping. If one takes Z to the left of X_n, then the lead angle is more than $+270°$ or $-90°$. Therefore the tripping criterion in terms of phase is

$$-90° < \frac{X_n - Z}{X_n} < +90° \quad \text{for trip}$$

Substituting $Z = V/I$ and multiplying numerator and denominator by I, we get

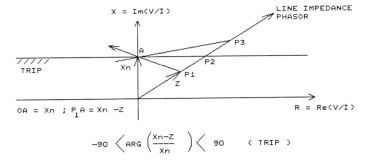

Figure 4.31 Reactance relay based on phase comparator.

$$-90° < \frac{IX_n - V}{IX_n} < +90° \quad \text{for trip}$$

This is implemented in Fig. 4.31.

Reactance relay based on amplitude comparator
Figure 4.32 shows the reactance characteristics with phasor $V/I = Z$ in the tripping, marginal and nontripping regions. OA represents the desired setting X_n of the reactance relay. Extend this phasor so that OB represents $2X_n$. The phasors are now

$P_1B = (2X_n - Z)$ (Z in trip region)

$P_2B = 2X_n - Z$ (Z for marginal operation)

$P_3B = 2X_n - Z$ (Z in nontrip region)

Note that $P_2B = OP_2$. Therefore, the tripping criterion for Z in the trip region is

$$|2X_n - Z| > |Z| \quad \text{(trip)}$$

Replacing Z by V/I and multiplying numerator and denominator by I, we get

$$\underbrace{|2IX_n - V|}_{\substack{\text{operating}\\\text{quantity}\\S_o}} > \underbrace{|V|}_{\substack{\text{restraining}\\\text{quantity}\\S_r}} \quad \text{(trip)}$$

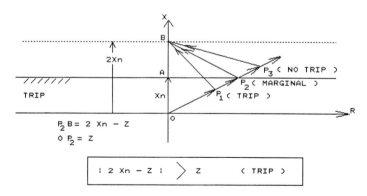

Figure 4.32 Reactance relay based on amplitude comparator.

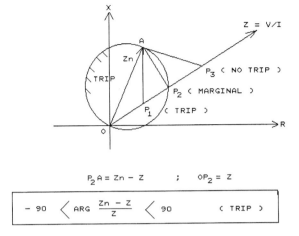

Figure 4.33 MHO relay based on phase comparator.

4.4.2.7. Static MHO Relay

Static MHO relay based on phase comparator
Figure 4.33 shows the MHO circle with $OA = Z_n$ as the diameter, and the three positions of impedance seen by relay $Z = V/I$ in the tripping region, threshold operation and the nontripping region. Note,

$P_1 A = Z_n - Z$ (for tripping)

$P_2 A = Z_n - Z$ (for marginal operation)

$P_3 A = Z_n - Z$ (for no tripping)

The tripping criterion is shown in the figure itself.

Static MHO relay based on amplitude comparator
Refer to Fig. 4.34. The proof is elementary and clearly shown in the figure.

Crosspolarization and memory
The conventional MHO relay, polarized from the fault voltage, suffers from the defect that when a fault occurs very close to the relay the fault voltage tends to vanish (i.e., it becomes virtually zero). Since the restraining and operating torques are directly proportional to the polarizing quantity, the relay fails to operate.

There are two methods to overcome this deficiency.

1. Tune the polarizing coil to power frequency (i.e., 50 or 60 Hz) so that the polarizing flux is maintained for a few cycles in spite of the fact the voltage collapses. During this time the relay operates

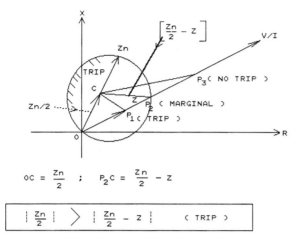

Figure 4.34 Static MHO relay based on amplitude comparator.

correctly. This method has been discarded because the supply frequency may vary due to a fault, whereas the polarizing coil resonates at a fixed frequency. The difference in the two frequencies, and hence the phase difference, increases with every cycle.

2. The second method is to use a *healthy* phase polarization, preferably cross-polarization.

Consider an example of a conventional A-E MHO unit, polarized from a fault voltage V_a, based on the phase comparator. The trip equation is

$$-90° < \text{Arg} \frac{(I_a + KI_0)Z_n - V_a}{V_a} < +90°$$

If V_a vanishes due to close-in fault, the comparator does not produce an output, so the relay fails to operate. The trick is to add the unfaulted voltage, which does not collapse much, to the fault voltage. The modified trip equation, with healthy phase cross-polarization (i.e., V_{bc}), is

$$-90° < \text{Arg} \frac{(I_a + KI_0)Z_n - \left|(1-k)V_a + \dfrac{kV_{bc}\angle 90°}{\sqrt{3}}\right|}{\left|(1-k)V_a + \dfrac{kV_{bc}\angle 90°}{\sqrt{3}}\right|} < 90°$$

where

$$V_{\text{polarizing}} = (1 - k)V_a + \frac{kV_{bc}\angle 90°}{\sqrt{3}}$$

k = amount (fraction of V_{bc}) of polarization

If $k = 0$,

$V_{\text{polarizing}} = V_a$ (conventional relay)

If $k = 1$,

$$V_{\text{polarizing}} = \frac{V_{bc}\angle 90°}{\sqrt{3}} \rightarrow (100\% \text{ cross-polarization})$$

$= V_a$ under balanced conditions

$\neq V_a$ under fault conditions

This type of polarization gives the following results:

1. Under balanced conditions (i.e., three-phase fault and power swing), the relay characteristics remain similar to a conventional MHO and, therefore, practically immune to power swing.
2. Under fault conditions the relay positively operates, since the polarizing quantity does not vanish.

Extensive analytical studies have been made by Wedephol, whose paper has been cited in the references.

4.4.2.8. Static QDR

QDR based on phase comparator

Combination of straight lines. The simplest technique for synthesizing QDR is to have four straight-line characteristics on the *R-X* diagram with their trip outputs AND compounded (Fig. 4.35). If four straight-line relay characteristics can be synthesized by a suitable comparator and if their trip outputs are AND compounded (i.e., trip contacts in series), then the common area of the four characteristics is the resultant trip area. This is one method of designing the QDR. A disadvantage of this method is that failure of one comparator completely distorts the relay characteristics, thereby leading to malfunctions. This method is now rarely used.

As an example, let us find the inputs to a cosine comparator to develop straight-line characteristics. Refer to Fig. 4.36. Drop a perpendicular on the desired straight-line characteristics from the origin of the *R-X* diagram. Let this impedance be $OA = Z_n$. Assume Z in the tripping, marginal operation

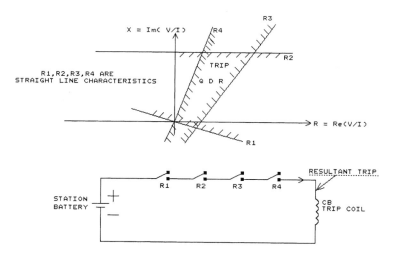

Figure 4.35 QDR relay as a combination of straight lines.

and nontripping regions to be OP_1, OP_2 and OP_3, respectively, in the first and second quadrants.

The phasors P_1A, P_2A, and P_3A are $Z_n - Z_r$, which depend on the magnitude of Z. The tripping criterion is now simply

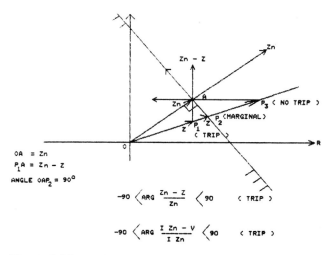

Figure 4.36 Straight-line characteristics.

Chapter 4

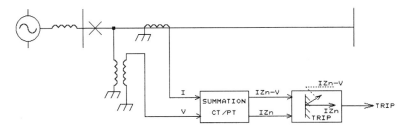

Figure 4.37 Block diagram to generate a straight line.

$$-90° < \text{Arg}\, \frac{Z_n - Z_r}{Z_n} < +90°$$

Substituting $Z_r = V/I$ and multiplying by I, we get

$$-90° < \text{Arg}\, \frac{IZ_n - V}{IZ_n} < +90°$$

Figure 4.37 shows the block diagram for generating the desired straight-line characteristics. The four comparators with their trip outputs are logically AND compounded.

QDR based on asymmetrical phase comparator. The technique is identical to that already described, except that one employs two asymmetrical comparators to generate a combination of two straight lines on the R-X diagram with one discontinuity. This is shown in Fig. 4.38. If the two trip outputs of the two comparators are logically AND compounded, we obtain the desired QDR.

The comparators required have asymmetrical phase angle margins, unlike the +90°/−90° for cosine comparators. Figure 4.39 shows the phase angle criterion.

$$\alpha < \text{Arg}\, \frac{S_m}{S_p} < \beta$$

Figure 4.39 also shows how the combination of two straight lines is synthesized. The inequality phase relationships for the two characteristics are

$$-\beta < \text{Arg}\, \frac{Z_r - Z_b}{-Z_b} < \alpha$$

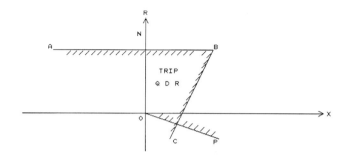

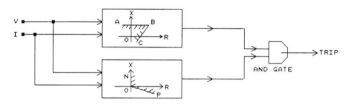

Figure 4.38 QDR based on two asymmetrical comparators.

and

$$90° + v > \text{Arg}\frac{Z}{Z_p} > 0$$

The hardware, based on integrated circuits, counters, etc., is beyond the scope of this book.

QDR based on multi-input phase comparator. Refer to Fig. 4.40, where points O, A, B and C completely define the QDR. Assume $V/I = Z$ in the tripping region as phasor OP. Join the point P_1, the tip of phasor Z, to points A, B and C. Notice that the four phasors of interest are

$$P_1O = -Z \qquad P_1B = Z_b - Z$$
$$P_1A = Z_a - Z \qquad P_1C = Z_c - Z$$

These phasors are confined to phase angles more than 180° (i.e., phase shift between any two adjacent phasors).

In the adjoining figure, let $V/I = Z$ be in the nontripping region. Note that the phasors P_2A, P_2O, P_2C, and P_2B are now confined to angles less than 180°. Note that this happens whenever the point P crosses any of the

Chapter 4

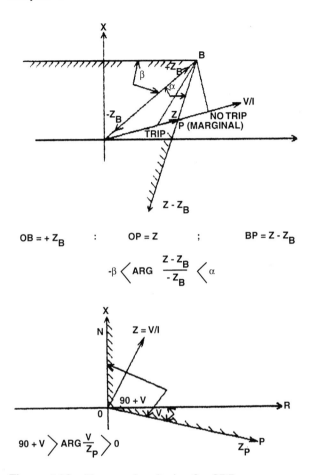

Figure 4.39 Phase angle criterion for QDR.

lines OA, AB, BC and CO to enter the nontripping region (Fig. 4.41), wherein it is shown that the positive coincidence period

1. is zero if the phasors are confined to a phase angle margin of more than 180°.
2. is finite if the phasors are confined to a phase angle margin less than 180°. The finite coincidence period is decided by only two extreme phasors.

To develop the QDR, we must check the CP of the four phasors; thus, if it is (a) zero → trip, (b) finite → no trip.

118 **Distance Protection**

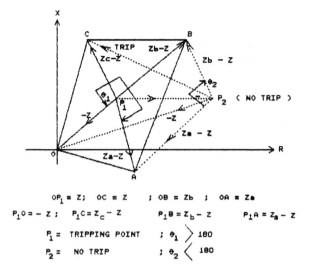

$OP_1 = Z; \quad OC = Z \quad ; \quad OB = Z_b \quad ; \quad OA = Z_a$

$P_1O = -Z; \quad P_1C = Z_c - Z \quad P_1B = Z_b - Z \quad P_1A = Z_a - Z$

$P_1 = $ TRIPPING POINT $\quad ; \quad \theta_1 > 180$

$P_2 = $ NO TRIP $\quad ; \quad \theta_2 < 180$

Figure 4.40 QDR based on multi-input phase comparator.

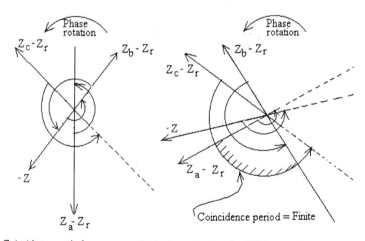

Coincidence period = zero · · Each phasor positive for 180 in counterclockwise direction

Figure 4.41 Coincidence period of more than two inputs.

Chapter 4

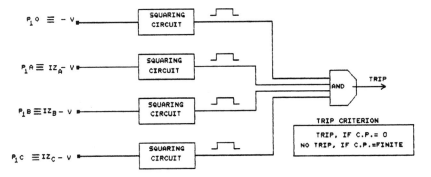

Figure 4.42 Block diagram of QDR based on multi-input phase comparator.

Figure 4.42 shows the block diagram of the relay.

QDR based on multi-input amplitude comparator
If a multi-input amplitude comparator has n sinusoidal inputs, it is equivalent to $n - 1$ two-input amplitude comparators with their trip outputs AND

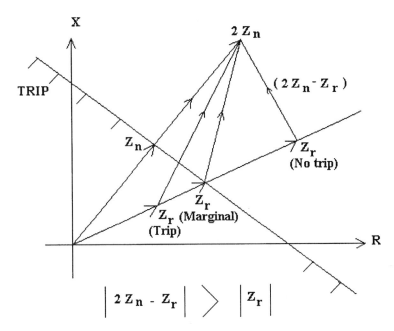

Figure 4.43 QDR-based multi-input amplitude comparator.

compounded. For example, let the circuitry satisfy the following four equations with five inputs (i.e., $n = 5$):

$(S_1 < S_2)$, $(S_1 < S_3)$, $(S_1 < S_4)$, and $(S_1 < S_5)$ for resultant trip.

As a result of logic AND compounding of the individual trip outputs, the relay characteristics will be the common area generated by individual amplitude comparators. Let us use this five-input amplitude comparator to generate the desired quadrilateral distance relay.

Figure 4.43 shows the application of amplitude comparator technique for generating a straight line. Such four comparators with their trip outputs and compounded generates a QDR.

4.4.3. Microprocessor-Based Distance Relays

A microprocessor is a smaller version of a digital computer due to the bit length of the data (i.e., 8 bit or 16 bit), the address bits (i.e., 16 bit or 32 bit) and the slower processing time (time required for arithmetic operations etc.). Otherwise it is no different from a computer.

The potential of the processor for its application to relaying is enormous, mainly due to its wide set of instructions, such as arithmetical, logical, branch, decision making, loop, memory, etc., and extremely small processing time.

There is a growing interest in developing processor-based protective relays which are more flexible, because they are programmable, and which are superior to conventional EM or static relays. The main features which have encouraged the design and development of processor-based relays are their economy, compactness, reliability, flexibility and improved performance.

Usually the relay voltage and current are full of harmonics and dc offset, which in static relays are filtered out by analog filters, which contain energy-storing elements such as capacitors and, perhaps, inductors. These elements, while performing the duty of filtering unwanted noise, produce their own secondary transient. To some extent, this defeats the purpose of analog filters in filtering the unwanted signal.

The processor is capable of filtering this noise by suitable algorithms (i.e., Fourier, Walsh, Harr, etc.) without producing secondary transients. In addition, the processor continuously monitors and stores in its memory the prefault and postfault conditions. These data are extremely useful in fault diagnosis.

The relay program is very complicated in machine language (i.e., op codes), and it is absolutely desirable to have the program in this machine language, since it is the fastest to execute. Microprocessor development systems (MDS) are available where one writes the program in a higher language, such as C, and cross-compiles in machine code. In this way programming difficulties can be overcome.

Another advantage of processor-based relays is the on-site changing of the relay characteristics. For example, in one overcurrent relay, one can choose definite time, inverse, very inverse, extremely inverse, etc., OC relay characteristics by a flick of the keypad.

The processor continuously monitors the status of the keypad and, depending on the changes in it, jumps to the desired program to generate the new characteristics. In distance relays it is possible to jump to different relay characteristics. This has virtually led to phasing out EM relays, and soon even static relays will be replaced.

4.4.3.1. Interfacing Relay Voltage and Current to Microprocessor

Figure 4.44 shows the general pattern of interfacing voltage, current, keypads, displays, etc. It needs no explanation, except that since the relay voltage inputs vary (current increases and voltage tends to drop on a fault) and are bipolar (i.e., positive and negative going), the analog-to-digital interface has to be bipolar and suitably protected.

4.4.3.2. Digital Calculation of RMS Values of Current, Voltage, and Impedance (Its Modulus and Angle)

Before we take up the design of various relay characteristics, recall that the basic inputs to the processor-based relay are analog quantities, such as sinusoidal voltages and currents, perhaps full of unwanted noise (harmonics and dc offset). Hence, filtering is required to extract only the fundamental frequency components, and then the RMS values, impedance = V_{rms}/I_{rms}, impedance angle θ_r, etc., must be calculated and a tripping or no-tripping decision must be made. Note that prefault and postfault inputs are available on the display for fault diagnosis.

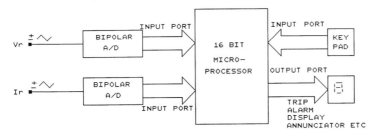

Figure 4.44 Interfacing to processor.

Calculation of fundamental frequency RMS values

Several methods have been published for the digital calculation of RMS values of sinusoidal currents and voltages using a processor. We shall choose the two most frequently used algorithms for their computational speed and ability to filter out harmonics and dc offsets.

Fourier technique (Discrete). In this method, the fundamental (50 or 60 Hz) sine and cosine components are obtained by correlating the incoming data samples of voltage or current (output of A/D converter) with the stored samples of the reference fundamental sine and cosine functions. The Fourier series expansion of any periodic signal $x(t)$, with period T (i.e., 20 msec for 50 Hz and 16.67 msec for 60 Hz) is

$v(t)$ = dc component + sine component of harmonics
 + cosine components of harmonics

$$x(t) = V_o + \sqrt{2}V_{1,s} \sin wt + \sqrt{2}V_{1,c} \cos wt \\ + \sqrt{2}V_{2,s} \sin 2wt + \sqrt{2}V_{2,c} \cos 2wt$$

$$= \frac{\text{dc} + \text{sine components}}{\text{up to } n\text{th harmonics}} + \frac{\text{cosine components}}{\text{up to } n\text{th harmonics}}$$

where

$$V_o = \int_{wt=0}^{wt=2\pi} v(wt)\, d(wt) \; \frac{1}{2\pi} \int_0^{2\pi} V(wt)\, d(wt)$$

= dc component

$$V_{1,s} = \int_{wt=0}^{wt=2\pi} v(wt) \sin wt\, d(wt) \; \frac{1}{\pi} \int_0^{2\pi} V(wt) \sin wt\, d(wt)$$

= fundamental sine component

$$V_{1,c} = \int_{wt=0}^{wt=2\pi} v(wt) \cos wt\, d(wt) \; \frac{1}{\pi} \int_0^{2\pi} V(wt) \cos(wt)\, d(wt)$$

= fundamental cosine component

$$V_{1,\text{RMS}} = \sqrt{V_{1,s}^2 + V_{1,c}^2}$$

Thus, the periodic samples taken of $v(wt)$ by the A/D converter are to be multiplied (called correlation) by stored weighting coefficients of the fundamental sine and cosine functions and then added to get $V_{1,s}$, $V_{1,c}$ and $V_{1,\text{RMS}}$.

Consider a numerical example with periodic sampling rate of 16 samples per full-cycle window of the highly distored voltage signal $v(t)$. The sampling instants for the input voltage $v(t)$ are every $360°/16 = 22.5°$. Then

Chapter 4

$$V_{1,s} = \frac{1}{16} \sum_{n=0}^{n=15} w_{n,s} v_n = \text{RMS sine component fundamental}$$

where

$v_0 - v_{15}$ = sampled values of $v(t)$
$w_{0,s} - w_{15,s}$ = weighting coefficients of fundamental sine function
   = sin 0°, sin 22.5°, sin 45°, etc.

$$V_{1,c} = \frac{1}{16} \sum_{n=0}^{n=15} w_{n,c} v_n = \text{RMS cosine component fundamental}$$

where

$w_{0,c} - w_{15,c}$ = weighting coefficients of fundamental cosine function
   = cos 0°, cos 22.5°, cos 45°, etc.

Figure 4.45 shows the sampling instants and weighting coefficients of fundamental sine and cosine correlating functions. Note that the weighting coefficients need to be precalculated and stored in the memory.

To extract the RMS value of the fundamental component from $v(t)$, the processing clock cycles, for an INTEL 8086 CPU, are

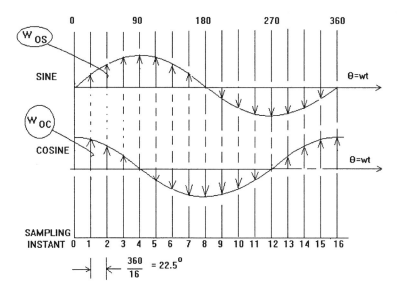

Figure 4.45 Weighting coefficients for Fourier series.

No. of adds./sub.	No. of mults.	Total clock cycles Adds./sub. = 3 Mults. = 1 28
32	32	32 × 3 + 32 × 128 = 4192

The frequency response is

DC	Funda.	2nd	3rd	4th	5th	7th	8th	Harmonics
0	100%	0	0	0	0	0	0	

The full-cycle window Fourier is the best, since it eliminates dc and all odd and even harmonics. Unfortunately, its computational time is quite high and therefore it is the slowest real-time algorithm. Faster computational times can be obtained by employing the Walsh algorithm.

Walsh algorithm. The Walsh correlating functions are square-wave type and therefore the weighting coefficients are either +1.0 or −1.0, unlike Fourier. This leads to no multiplications, and therefore the Walsh algorithm is computationally quite fast.

Whenever we intend to divert correlating functions from Fourier (which are purely sinusoidal) to any other function (i.e., Walsh square wave), we must understand that if the correlating function is other than sinusoidal, then the new correlating function will contain certain harmonics in a certain percentage of the fundamental component. Therefore after correlation, not only the fundamental component but also the harmonics will be extracted. This results in degraded frequency response at the cost of trying to improve computational speed.

For example, Fourier analysis of a square wave tells us that it can be reconstituted by adding a certain percentage of odd harmonics to the fundamental component. The equation is

$$W_{1,s} = \sin wt + \frac{1}{3} \sin 3wt + \frac{1}{5} \sin 5wt + \frac{1}{7} \sin 7wt + \text{higher terms}$$

Chapter 4

This will be clarified as we take the example of a fundamental Walsh correlating function. Consider a sampling rate of 16 samples per cycle. Then

$$\frac{1}{16}\left[\sum_{0}^{7} V_n(+1) + \sum_{8}^{15} V_n(-1)\right] \quad \text{(RMS funda. sine comp.)}$$

where

$v_0 - v_{15}$ = sampled values of $v(t)$
$+1, -1$ = weighting coefficients of fundamental sine Walsh function

and

$$V_{1,c} = \frac{1}{16}\left[\sum_{0}^{3} V_n(+1) + \sum_{4}^{11} V_n(-1) + \sum_{12}^{15} V_n(+1)\right]$$

(RMS fund. cosine comp.)

where $+1, -1$ = weighting coefficients of fundamental cosine Walsh function.

If for brevity we assume that the relay input voltage has fundamental and odd harmonics, then after correlation with the fundamental sine Walsh function, the output is

$$\int v(wt)(\text{Walsh}_{\sin}) \, d(wt) = F_1 + \left(\frac{1}{3}\right) F_3 + \frac{1}{5} F_5 + \cdots$$

where F_1, F_3, F_5 are the RMS values of the fundamental, third, fifth, and higher odd harmonics, constituting input relay voltage $v(wt)$. Thus, after correlation, instead of completely filtering out all odd harmonics, a certain percentage is retained. Therefore the fundamental Walsh functions have an extremely poor frequency response, since the Walsh function itself contains harmonics. In effect, we correlate the input function $v(wt)$ with the fundamental sinusoid and with the harmonics constituting the square wave.

To extract the fundamental only, one must correlate the input function $v(wt)$ with higher-order (i.e., third, fifth, seventh, etc.) Walsh functions and extract a portion of their output from the output, when correlated with the fundamental Walsh. Without going into detail it has been proved that

$$F_1 = 0.9W_{1,s} - 0.373W_{3,s} - 0.074W_{5,s} - 0.179W_{7,s}, \text{ etc.}$$

where F_1 = desired fundamental sinusoidal component of $v(wt)$
$W_{1,s}$ = component of $v(wt)$ extracted by correlating with fundamental sine Walsh
$W_{3,s}$ = component of $v(wt)$ extracted by correlating with third harmonic of sine Walsh

$W_{5,s}$ = component of $v(wt)$ extracted by correlating with fifth harmonic of sine Walsh

$W_{7,s}$ = component of $v(wt)$ extracted by correlating with seventh harmonic of sine Walsh

The Walsh algorithm therefore slows down computationally, due to multiplications by 0.9, 0.373, 0.074, 0.179, etc. The fundamental, third, and fifth sine Walsh correlating functions are shown in Fig. 4.46.

The following table shows the frequency response and computational time for extracting sine and cosine components from $v(wt)$.

Computational Time for Sine/Cosine

Algorithm	No. of adds./subs.	No. of mults.	Total clock cycles Add/sub = 3 cycles Mults. = 128 cycles
Walsh (one term)	32	nil	32 × 3 = 96
Walsh (two terms)	32	4	32 × 3 + 4 × 128 = 608

Frequency Response

Algorithm	Processing time (clock cycles)	Fund.	3rd	5th	7th
Walsh one term	96	100%	33.3%	20%	14.3%
Walsh two terms	608	100%	3%	0%	20.0%

The two-term Walsh does eliminate the third and fifth harmonics but it cannot get rid of the seventh harmonic in $v(wt)$. An additional third term Walsh is required to eliminate it, but this function introduces additional multiplications, thereby increasing the computational time and reducing the sampling rate.

Calculation of impedance. The inputs to the processor are basically distorted relay voltage and relay current. One has to extract only the fundamental components to find the impedance. To extract the fundamental, choose one of the algorithms from among Fourier, Walsh, Harr, etc. The

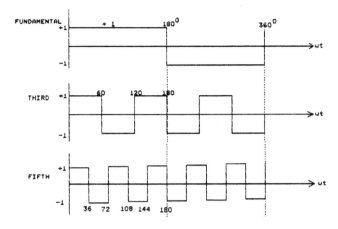

Figure 4.46 Walsh functions (sine).

choice depends on the processing time of the algorithm and the highest harmonic to be filtered out. For example, assume that we have 16 samples each of $v(wt)$ and $i(wt)$. We have to find the sine and cosine components of voltage and current and the phase angle, and then determine the impedance.

This process is continuing in the sense that if after computation you find that there is no fault, then take only one new sample of $v(wt)$ and $i(wt)$, dispose of the oldest sample, and recalculate the new impedance.

Let the processing time between adjacent samples be T_p. Then

$$\text{Sampling rate} = \frac{\text{Period of fundamental}}{T_p} \text{ samples/cycle}$$

As a numerical example let

T_p = processing time = 1.0 msec

T = period of fundamental = 20 msec   (on 50-Hz base)

       = 16.67 msec   (on 60-Hz base)

Then

Samples/cycle ≤ 20   (on 50-Hz base)

Samples/cycle ≤ 16   (on 60-Hz base)

By the Nyquist criterion the highest harmonic that can be filtered out is half the sampling rate. The highest odd harmonic that can be filtered out with the above sampling rate is

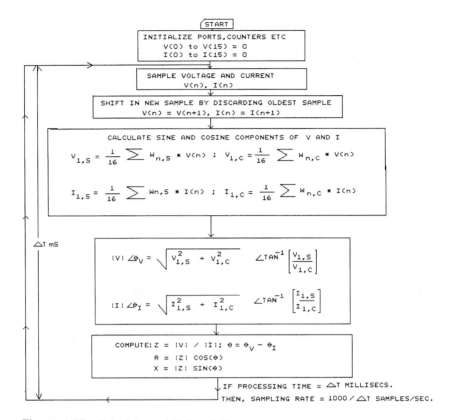

Figure 4.47 Calculation of Z, R and X.

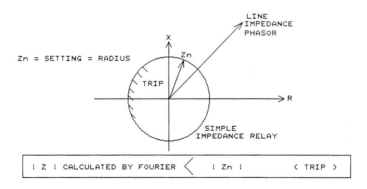

Figure 4.48 Simple impedance relay.

Chapter 4

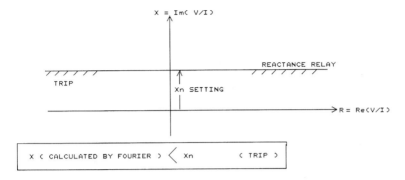

Figure 4.49 Reactance relay.

Highest harmonic that can be eliminated = ninth (for 50 Hz)
Highest harmonic that can be eliminated = seventh (for 60 Hz)

The flowchart for calculating impedance Z, resistance R and reactance X is shown in Fig. 4.47

Various relay characteristics. (i) Simple impedance relay: Refer to Fig. 4.48. It is the simplest relay, and trip output is to be issued if $|Z| \leq$ setting Z_n (i.e., radius of SIR).
(ii) Reactance relay: Refer to Fig. 4.49.

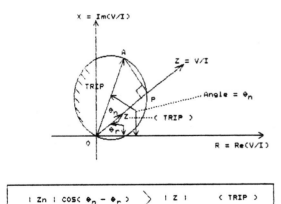

Figure 4.50 MHO relay.

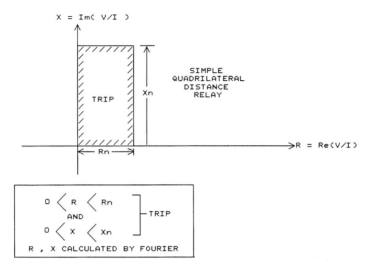

Figure 4.51 The QDR.

$X \leq X_n$ (trip)

(iii) MHO relay: Refer to Fig. 4.50, which shows the MHO characteristics.

$$|Z_n \cos(\theta_n - \theta_r)| \geq |Z_r|$$

(iv) Quadrilateral: Refer to Fig. 4.51, showing the QDR on the R-X diagram. If $0 < R < R_n$ and $0 < X < X_n$, then trip output occurs.

4.5. EFFECT OF POWER SWING ON DISTANCE RELAYS

4.5.1. Stable and Unstable Power Swings

The purpose of protective relays is to detect the fault and trip the appropriate CB; they should not operate for conditions other than fault. If they do, such operations are called malfunctions. We shall study the effect of power swing on distance relays. Refer to Fig. 4.52, which shows a tie line connecting two synchronous systems.

Sources E_A and E_B have impedances Z_{sA} and Z_{sB}, respectively. Source E_a leads source E_b by an angle δ. If δ is positive, power transfer is from source A to source B. If δ is negative, power transfer is from source B to source A. The angle δ is called the *power angle, rotor angle* or *load angle*. The tie line A-B has Z_L as its line impedance. There are distance relays R_A

Chapter 4

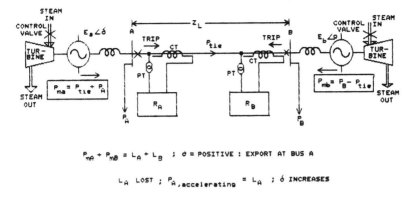

Figure 4.52 Tie line connecting two synchronous systems.

and R_B, along with CBs, at both ends of the line. The purpose of these relays is to operate on a fault, in the trip direction, on the line. They have no other purpose.

Consider local loads P_A and P_B on buses A and B, respectively. Let the steady-state power flow on the tie line be P_{tie}. The mechanical power inputs (i.e., turbine power), under steady-state conditions and with all losses neglected, to the two machines are

$$P_{A,mech} = P_A + P_{tie} \qquad P_{B,mech} = P_B$$

Now assume that the local load on bus A is lost. Note that there is no fault on the tie line, so relays R_A and R_B should not operate. The loss of local load P_A produces acceleration of machine A because of the mismatch between the mechanical input and electrical output.

$$P_{acc,A} = P_{mech} - P_{tie} = [P_A + P_{tie}] - P_{tic} = P_A$$

This results in the following action:

1. Machine A will accelerate, causing increase in power angle δ.
2. P_{tie} increases due to increase in rotor angle.
3. Speed governor of machine A will sense the increase in speed and will act to reduce the mechanical input by closing the steam valve.
4. If the load disturbance P_A is small, the changes in δ will be small, so the system will be ultimately (i.e., steady state) stabilized by the speed-governing control system. Such swinging of δ is called *stable power swing*.
5. If P_A is large, the changes in δ will be much greater and produce instability. Angle δ will continue to increase.

The power angle versus time for stable and unstable power swing appears somewhat as shown in Fig. 4.53. Since the line is not faulted and the power disturbance is outside the line, and if resulting power swing is going to be stable, there should be no operation of relays R_A and R_B, unless there is a malfunction.

The ultimate situation is that

1. Loss of relatively small load outside the line causes stable power swing, from which the system is capable of recovering (i.e., due to speed-governing control mechanism and perhaps the automatic voltage regulator).
2. Temporary increase in load angle δ causes increase in tie-line current and reduction in voltage at the relay location.
3. The impedance seen by the relay, on increased angle δ, will drop, and if this impedance, which is a function of δ, gets into the trip zone of the relay the relay will malfunction on the stable power swing. This malfunction causes further disturbance in the system and, in a multimachine system, results in cascade tripping of many more relays and CBs. There is a total collapse of the integrated system, causing great concern in terms of load dislocation and time-consuming resynchronization of the whole system.
4. Thus, one has to investigate the impedance seen by the distance relay as a function of δ.

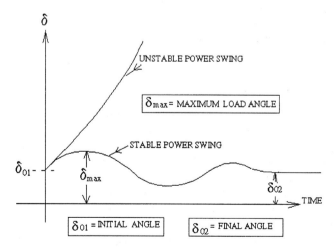

Figure 4.53 Stable and unstable power swing.

4.5.2. Impedance Seen by Relay on Power Swing

Refer to Fig. 4.54. For simplicity we neglect the load impedances, so the voltage and current as functions of δ for R_A are

$$I_r = \frac{E_A \angle \delta - E_B \angle 0}{Z_t}$$

$$Z_t = Z_{sA} + Z_l + Z_{sB}$$

$$V_r = E_A \angle \delta - I_r Z_{sA}$$

We can rearrange the equations in terms of impedances alone:

$$Z_t + \frac{E_B \angle 0}{I_r} - \frac{E_A \angle \delta}{I_r} = 0 \tag{1}$$

and

$$Z_r = -Z_{sA} + \frac{E_A \angle \delta}{I_r} = f(\delta) \tag{2}$$

These equations give us a graphical method, which is much more revealing than the analytical one, for constructing impedance triangles and locating Z. Figure 4.55 shows the construction. The method is as follows for relay R_A.

1. Draw an R-X diagram. Relay R_A is located at point O, the origin of the R-X diagram.
2. Draw the impedance behind relay location Z_{sA} in a negative direction from the origin: $OA = -Z_{sA}$.

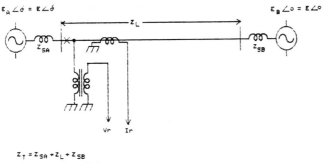

Figure 4.54 Impedance seen by relay on power swing.

Distance Protection

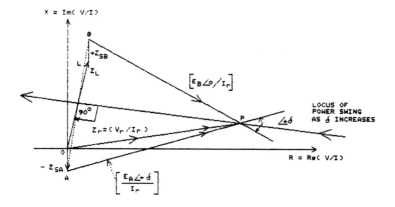

Figure 4.55 Locus of power swing construction.

3. Draw all the impedances forward of the relay location in a positive direction: $OL = +Z_L$ and $LB = +Z_{sB}$.
4. Join point A to point B. $AB = Z_t$.
5. Assume for simplicity that $|E_A| = |E_B| = |E|$.
6. Because the voltages are assumed to be equal, draw a perpendicular bisection of Z_t.
7. Take any point P on this bisector. Note that
   a. Impedance triangle ABP satisfies Eq. (1).
   b. Eq. (2) tells us that $Z_r = OP$.
8. The angle between phasor AP and BP is the power angle δ, which increases as P moves toward the line impedance. Therefore, the bisector is called the *locus of the power swing*.
9. If $|E_A| \neq |E_B|$, so that $E_A/E_B = n$, then the locus is a circle, as shown in Fig. 4.56 with

$$\text{Distance from point } B \text{ to center of power swing circle } C = \frac{Z_T}{n^2 - 1}$$

Chapter 4

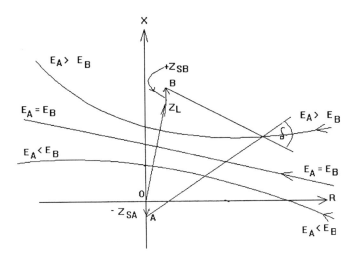

Figure 4.56 Locus of power swing when $E_A/E_B = n$.

and

$$\text{Radius of power swing circle} = \frac{nZ_t}{n^2 - 1}$$

4.6. DISTANCE SCHEME FOR THREE-PHASE LINES

To some extent, we have been inaccurate by saying that the distance relay, energized by voltage and current, measures the impedance from the relay location up to the fault point. We have drawn a line diagram of the three-phase transmission line and making a fault.

From Fig. 4.57 one has to realize that

1. At the relay location we have three phase-to-neutral voltages and three line-to-line voltages.
2. Further, we have three line currents and three difference in line currents at the relay location.
3. At the fault point we may have a three-phase, L-L-G, L-L or L-G fault.
4. The transmission line has positive, negative and zero-sequence impedances from the relay location up to the fault point. Note that all three-phase static equipment, such as transformers and lines, have

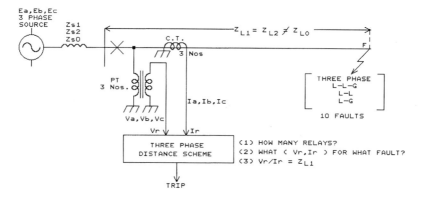

Figure 4.57 Three-phase line: voltages, currents and faults.

$$Z_{\text{positive}} = Z_{\text{negative}} \neq Z_0$$

This assumption is not true for rotating equipment such as turboalternators, induction motors, etc. Even then, for physical modeling, we assume that

$$Z_{\text{positive,source}} = Z_{\text{negative,source}} = Z_{0,\text{source}}$$

Therefore, the moot question is

Among the three line impedances (i.e., positive, negative or zero sequence) which impedance should the distance relay measure? Notice that positive sequence quantities (positive sequence voltage and current) are the only ones in all 10 faults (i.e., three-phase, L-L-G, L-L and L-G faults). The distance relay, therefore, is so energized that it measures the positive sequence impedance of the faulted line.

The next question is, which voltage and current must be chosen for a particular fault so that the ratio of V/I or the impedance seen by the relay is only positive sequence impedance.

These questions have been solved and the answer is

Three distance relays are required to locate the seven phase faults (i.e., three-phase, L-L-G and L-L). The relays are called *phase-fault measuring units* (i.e., distance relays) and are required to be energized by line-to-line voltages and difference in line currents, so that they measure the positive sequence impedance.

Further, for L-G faults, the voltage and current to the ground-fault measuring units are found to be phase voltage and phase current, suitably compensated by the zero-sequence current. This choice ensures that only the positive sequence impedance is measured on the faulted line.

Chapter 4

Thus, to protect a three-phase line against all 10 faults, we need six measuring units (distance relays). Of these six, three are called phase-fault measuring units and three are called ground-fault measuring units.

4.6.1. Phase-Fault Measuring Unit

Consider a B-C fault on the line as shown in Fig. 4.58. We want to determine V_r and I_r so that

$$\frac{V_r}{I_r} = [Z_{\text{positive}}]^{\text{B-C fault}}$$

where Z_{positive} = positive sequence impedance of line from relay location up to fault point. The solution is

$$V_r = V_{BC} \qquad I_r = I_B - I_C$$

so that

$$\frac{V_{BC}}{I_B - I_C} = Z_{\text{positive}}$$

There are two proofs for this; one is the physical interpretation, which is easy to comprehend, and the other uses symmetrical components.

(a) Physical interpretation: Figure 4.59 shows a three-phase generator, source impedance Z_s, the B-C relay location and the three-phase line with B-C fault. This physical model, whose proof is left to the reader, ensures that $Z_{\text{positive}} = Z_{\text{negative}} \neq Z_0$ for the line as well as the source. Note that

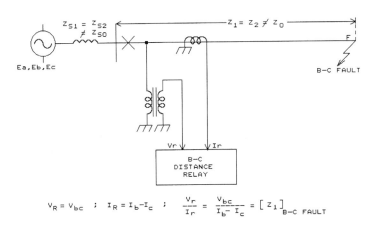

Figure 4.58 Three-phase line with B-C fault.

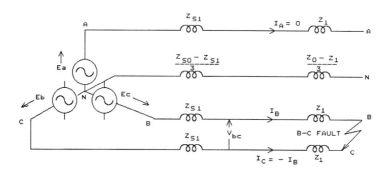

Figure 4.59 Physical model for choice of voltage and current for B-C measuring unit.

$$I_A = 0 \quad \text{and} \quad I_B = -I_C$$

By Kirchhof's law

$$\frac{V_{BC}}{I_B - I_C} = Z_{\text{positive}}$$

(b) By symmetrical components: By symmetrical component theory

$$V_{BC} = V_B - V_C = [a^2 - a][V_{A1} - V_{A2}]$$
$$I_B - I_C = [a^2 - a][I_{A1} - I_{A2}]$$
$$\frac{V_{BC}}{I_B - I_C} = \frac{V_{A1} - V_{A2}}{I_{A1} - I_{A2}} \text{ irrespective of fault}$$

The three sequence networks of phase A, the relay location and their connections for a B-C fault are shown in Fig. 4.60.

By applying Kirchhoff's law to the closed circuit between the two sequence networks, we have

$$\frac{V_{A1} - V_{A2}}{I_{A1} - I_{A2}} = \frac{V_{BC}}{I_B - I_C} = Z_1 \quad \text{for B-C fault}$$

The next question is, how does the B-C measuring unit behave for

Chapter 4

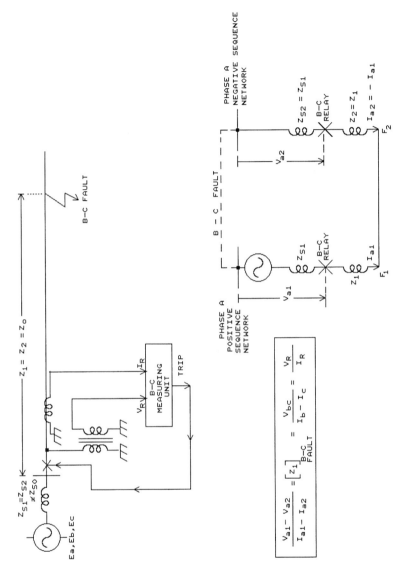

Figure 4.60 B-C fault/sequence networks.

faults other than B-C. These investigations are normally called the performance of healthy phase relays and have been extensively reported. We shall take those cases in which we are profoundly interested.

(i) B-C measuring unit on B-C-G fault: The phase A sequence network connections for this fault is shown in Fig. 4.61. It is easy to prove that

$$\frac{V_{A1} - V_{A2}}{I_{A1} - I_{A2}} = \frac{V_{BC}}{I_B - I_C} = Z_1 \quad \text{for B-C-G fault}$$

(ii) B-C unit on three-phase fault: For a three-phase fault, there is only a positive sequence network connected as in Fig. 4.62. In this case

$$V_{A2} = 0, \quad I_{A2} = 0 \quad \frac{V_{A1}}{I_{A1}} = Z_1$$

Therefore,

$$\frac{V_{A1} - V_{A2}}{I_{A1} - I_{A2}} = \frac{V_{A1}}{I_{A1}} = Z_1 \quad \text{for three-phase fault}$$

The B-C measuring unit, therefore, correctly measures the positive sequence impedance of the faulted line for a B-C fault, a B-C-G fault and a three-phase fault.

It is unnecessary to prove that the inputs to A-B and C-A measuring units will be

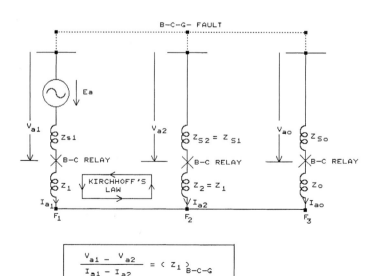

Figure 4.61 Sequence networks for B-C-G fault.

Chapter 4

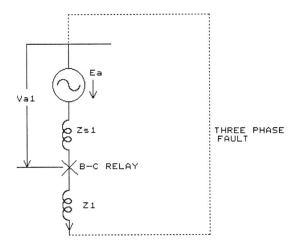

Figure 4.62 Sequence networks for three-phase fault.

$$V_r = V_{AB}, \quad I_r = I_A - I_B \quad \text{for A-B unit}$$
$$V_r = V_{CA}, \quad I_r = I_C - I_A \quad \text{for C-A unit}$$

The results of three-phase fault measuring units are summarized in the following chart.

Measuring unit	V_r	I_r	Impedance seen for various faults
A-B	V_{AB}	$I_A - I_B$	$= Z_1$ for A-B, A-B-G and three-phase $\neq [Z_1]$ for remaining
B-C	V_{BC}	$I_B - I_C$	$= Z_1$ for B-C, B-C-G and three-phase $\neq [Z_1]$ for remaining
C-A	V_{CA}	$I_C - I_A$	$= Z_1$ for C-A, C-A-G and three-phase $\neq [Z_1]$ for remaining

4.6.2. Ground-Fault Distance Relay

Here we ask, what relay voltage and what relay current can be fed to the A-E distance relay so that

Distance Protection

$$\frac{V_r}{I_r} = Z_1 \quad \text{for A-E fault}$$

(a) Physical interpretation: Figure 4.63 shows the physical model of a three-phase system, with A-E relay location and the A-E fault on the line. By applying Kirchhoff's law to the closed loop of the faulted circuit, we have

$$V_A = I_A Z_1 + I_0(Z_0 - Z_1)$$

Factoring Z_1, we have

$$V_A = Z_1 \left(I_A + I_0 \frac{Z_0 - Z_1}{Z_1} \right)$$

or

$$Z_1 = \frac{V_A}{I_A + kI_0} = \frac{V_r}{I_r}$$

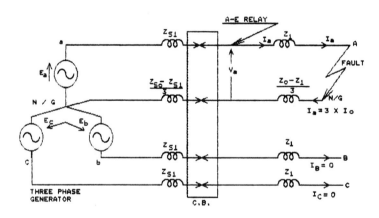

Figure 4.63 Physical interpretation of A-E fault.

Chapter 4

where

$$k = \frac{Z_0 - Z_1}{Z_1}$$

The equation says that for an A-E fault

$$V_r = V_A \quad \text{and} \quad I_r = I_A + kI_0$$

where

$$k = \frac{Z_0 - Z_1}{Z_1}$$

= zero-sequence current compensation factor, a fixed parameter of line

Thus, we say that the A-E ground-fault relay is to be energized by the phase voltage (i.e., V_A) and phase current suitably compensated by the zero-sequence current (i.e., $I_A + kI_0$).

(b) By symmetrical components: Figure 4.64 shows the phase A sequence networks connected in series to simulate an A-G fault. Applying Kirchhoff's law to the closed loop gives

$$(V_{A1} - I_{A1}Z_1) + (V_{A2} - I_{A2}Z_1) + (V_{A0} - I_{A0}Z_0) = 0$$

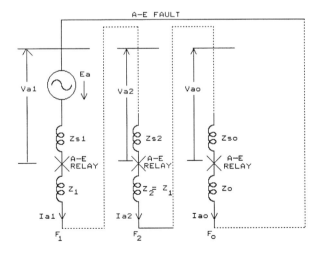

$$V_{a1} + V_{a2} + V_{a0} = I_{a1} Z_1 + I_{a2} Z_1 + I_{a0} Z_1$$

Figure 4.64 Sequence networks for A-E faults.

or

$$V_{A1} + V_{A2} + V_{A0} = I_{A1}Z_1 + I_{A2}Z_1 + I_{A0}Z_0$$

Adding and subtracting $I_{A0}Z_1$ on the right-hand side gives

$$V_A = I_{A1}Z_1 + I_{A2}Z_1 + I_{A0}Z_1 - I_{A0}Z_1 + I_{A0}Z_0$$
$$= Z_1(I_A + kI_{A0})$$

or

$$Z_1 = \frac{V_A}{I_A + kI_{A0}} = \frac{V_r}{I_r}$$

where

$$k = \frac{Z_0 - Z_1}{Z_1}$$

The A-E relay, therefore, is to be energized by phase A voltage and phase A current, suitably compensated by the zero-sequence current so that it correctly measures the positive sequence impedance.

The A-E relay performance for faults other than A-E has been thoroughly investigated, but we are more concerned with its performance for three-phase faults.

For a three-phase fault the sequence network connections are shown in Fig. 4.65. Note that $V_{A1}/I_{A1} = Z_1$, $V_{A2} = 0$, $V_{A0} = 0$, $I_{A2} = 0$, $I_{A0} = 0$ and $I_A = I_{A1}$. Thus, for an A-E relay,

$$\frac{V_r}{I_r} = \frac{V_A}{I_A + kI_{A0}} = \frac{V_{A1} + V_{A2} + V_{A0}}{I_{A1} + kI_{A0}} = \frac{V_{A1}}{I_{A1}} = Z_1$$

The A-E relay, therefore, measures correctly the positive sequence impedance of the faulted line for a three-phase fault. We require three ground-fault relays to take care of all three ground faults. Their inputs and the impedances seen are summarized as follows.

Measuring unit	Relay voltage	Relay current	Impedances seen on various faults
A-E	V_A	$I_A + kI_{A0}$	$= Z_1$ for A-E and 3-phase $\neq Z_1$ for other
B-E	V_B	$I_B + kI_{B0}$	$= Z_1$ for B-E and 3-phase $\neq Z_1$ for other
C-E	V_C	$I_C + kI_{C0}$	$= Z_1$ for C-E and 3-phase $\neq Z_1$ for other

Chapter 4

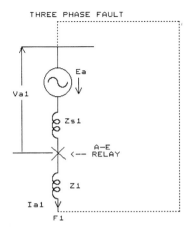

$$\frac{V_R}{I_R} = \frac{V_a}{I_a + K\ I_o} = \frac{V_{a1} + V_{a2} + V_{a0}}{I_{a1} + I_{a2} + I_{a0} + K\ I_{ao}} = \frac{V_{a1}}{I_{a1}} = \left[Z_1 \right]_{3\ \emptyset\ FAULT}$$

Figure 4.65 Sequence network connections for three-phase fault.

The total solution for protecting the three-phase line against all 10 faults is to have three phase-fault measuring units and three ground-fault measuring units. These six measuring units will operate on a three-phase fault correctly. From the operation of any one or more of them a guess can be made as to the type of fault.

4.6.3. Nonswitched Distance Scheme

Switching in a distance scheme has two interpretations:

1. As pointed out earlier, we need three phase-fault and three ground-fault distance relays, called measuring units. In the nonswitched distance scheme, pressure circuits and current circuits are permanently energized. Thus, there is no switching in or switching out of these basic relay quantities.

To generate a three-stepped distance scheme, we normally use the six measuring units for the first high-speed zone (i.e., Z_1, instantaneous). For generating a second zone, Z_2 with delay of T_1, and third zone, Z_3 with a further delay of T_2, the normal practice is to employ what is called *zone switching*. There should be no confusion in *zone switching* and switching of relay quantities.

One may use, cost permitting, 18 measuring units, 6 each for the three zones. If the cost is prohibitive, one may use six measuring units connected for Z_1 and have them zone-switched to Z_2 by a time delay T_1, and further switched to Z_3 by a further time delay T_2.

So, the nonswitched distance scheme has no switching of relay inputs, but may or may not have zone switching. The difference in nonswitched and switched distance scheme basically lies in whether you resort to switching relay inputs or not. The definitions, therefore, are

2. Nonswitched scheme: There is no switching of relay voltage and current to measuring units. There may or may not be zone switching. Either 6 or 18 measuring units are required. These schemes are applied to important HV/EHV lines, where one needs high-speed first zone protection.

If one needs zone switching, additional fault detectors are required to start the timer. The timer provides the required delay T_1 for Z_2 and T_2 for Z_3.

The trip scheme is shown in Fig. 4.66, whereas the settings (reach and operating time) of the three zones and fault detector are shown in Fig. 4.67. From the figure we note:

1. F_d is the fault detector. Its operating time is instantaneous and its reach is beyond zone Z_3.
2. The timer is a device, initiated by the fault detector and has two normally open contacts T_1 and T_2. The timer will close contact T_1 after a time delay of T_1, and the second pair of contacts T_2 will close after a time delay of T_2.
3. Z_1, Z_2 and Z_3 are the contacts of the three zone relays. Their operating time is instantaneous (i.e., without intentional time delay).
4. There is a station battery, normally 110 V dc, and the trip coil of the line circuit breaker.

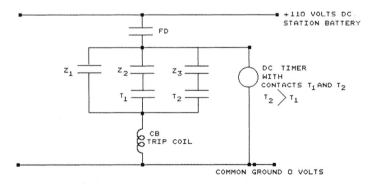

Figure 4.66 Trip scheme for nonswitched distance scheme.

Chapter 4

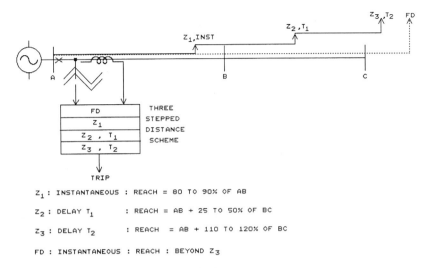

Z_1 : INSTANTANEOUS : REACH = 80 TO 90% OF AB

Z_2 : DELAY T_1 : REACH = AB + 25 TO 50% OF BC

Z_3 : DELAY T_2 : REACH = AB + 110 TO 120% OF BC

FD : INSTANTANEOUS : REACH : BEYOND Z_3

Figure 4.67 Setting zones and fault detectors.

Let us now look at the operation of the fault scheme in various zones.

(a) Fault in zone 1: This fault operates fault detector F_d and zone relays Z_1, Z_2 and Z_3. The CB, due to closure of F_d and Z_1, trips instantaneously, giving high-speed protection to faults within first zone.

(b) Fault in zone 2: This fault operates F_d, initiating the timer and Z_2 and Z_3. The fault, being beyond zone 1, does not operate Z_1. The timer closes its first pair of contacts T_1 after the preset delay of time T_1. The circuit breaker trips due to closure of F_d, Z_2 and T_1 with the desired second zone delay of T_1 sec.

(c) Faults in zone 3: This fault operates F_d, initiating the timer, and Z_3. The fault, being beyond zone 1 and zone 2, *does not* operate Z_1 and Z_2. The timer closes, first, contacts T_1, but since Z_2 is open there is no CB tripping. The timer continues to move, closing contacts T_2 after a delay T_2. The CB trips due to closure of F_d, Z_3 and T_2. The third zone fault is, therefore, cleared with a delay of T_2 sec.

4.6.4. Switched Distance Scheme

As the name implies, there is switching of relay inputs to the measuring units. This is over and above the required zone switching. The switching of relay inputs V_r and I_r, for different faults, is needed to reduce the measuring units from six to a more affordable number.

148   **Distance Protection**

Thus, the switched distance scheme has less than six measuring units, reducing the cost of protection at, as we shall see, the cost of time delay. It is pertinent to ask if a high-speed nonswitched distance scheme will be used, irrespective of line voltage (e.g., the same high-speed scheme for a 66-kV line and a 440-kV line. The American practice is "yes," whereas British and Continental practice is "no." British and Continental practice appears reasonable from the cost point of view. A switched distance scheme is cheaper than the nonswitched distance scheme.

Various switched schemes have been reported in the literature. We shall discuss only one type, which employs only one measuring unit for all seven phase faults. The reader can study other schemes. For the basic concept of switching "in" and "out" of relay voltage and relay current for different faults, discussing only one scheme is adequate. Figure 4.68 shows how one measuring unit, rather than three, can protect against all phase faults along

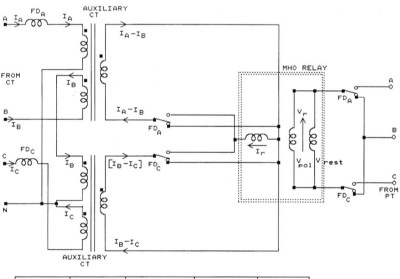

Figure 4.68 Switched phase-fault distance scheme.

Chapter 4

with the price. The price is in terms of slight delay in detecting the faults in the first zone itself. The protection engineer has to decide whether he or she can afford this delay.

The following chart shows the type of schemes and their approximate operating times for first-zone faults.

Type of scheme	No. measuring units	Minimum tripping time (cycles)	
		Phase faults	Ground faults
Nonswitched	6	1	1
Delta-wye	3	1	5
Interphase	2	5	5
Complete switching	1	5	5
Polyphase relays	1	1	1

In a delta-wye scheme the measuring units are reduced from 6 to 2. These three units are connected for phase faults. In the event of ground faults, the three units are energized, with the help of auxiliary relays, for ground faults (i.e., from L-L voltage and difference in line current to phase voltage and phase current suitably compensated for zero-sequence current). This scheme, therefore, needs additional fault detectors to detect whether the fault is phase fault or ground fault. The phase selectors are either overcurrent relays or MHO starters. A numerical example will later be given regarding their choice.

Since the MUs are connected for phase faults, the operating time is around 1 cycle. For ground faults the operating time increases to around 5 cycles, due to a delay of 4 cycles required to switch out the relay inputs for phase faults and to switch in the new inputs required for ground faults.

The interphase scheme employs one measuring unit for phase faults and one for ground faults. None of these units are energized by any voltage or current. Additionally, one needs fault selectors to detect the type of fault, and these phase selectors, with the help of auxiliary relays, switch in the correct voltage and current, depending upon the fault.

Although the measuring units operate in about 1 cycle, an additional delay of around 4 cycles takes place in the operation of the fault selectors and the auxiliary relays. Therefore the operating time for all faults increases to around 5 cycles.

In complete switching, the measuring units are reduced to the minimum possible, one. In this case the fault selectors have to switch in correct voltage and current for the type of fault detected by the fault selector.

4.6.5. Polyphase Distance Relay

The correct definition of the polyphase distance relay is: a polyphase distance relay consists of only one measuring unit, so energized that it correctly measures the positive sequence impedance of the faulted line, irrespective of fault type. Several attempts have been made to develop polyphase distance relays, but they have not been accepted in the field, due to variation of its characteristics on R-X diagram for different faults. An excellent attempt has been documented to develop such a relay, using a multi-input/multioutput pulse sequential circuit. See the paper listed in the references.

4.7. COMPARISON OF VARIOUS RELAY CHARACTERISTICS

The conventional distance relays are

1. Simple impedance
2. Reactance
3. MHO
4. Quadrilateral

To select proper characteristics for a line, we need to compare their performances in three areas:

1. Directional feature
2. Underreaching on arcing faults
3. Stability on power swing

4.7.1. Directional Feature

It has been proved that the impedance seen by a distance relay on a reverse fault is $-Z_r$, where Z_r is the impedance from the relay location to the reverse fault point. The minus sign is due to reversal of the relay input current on a reverse fault on double-end-feed lines.

Figure 4.69 shows all four relay characteristics, set to same reach along the line impedance phasor. The impedance $-Z_r$ for a reverse fault is shown in quadrant III. It can now be inferred that

1. SIR and reactance relays are nondirectional. They malfunction on reverse faults.
2. MHO and QDR are inherently directional.

Chapter 4

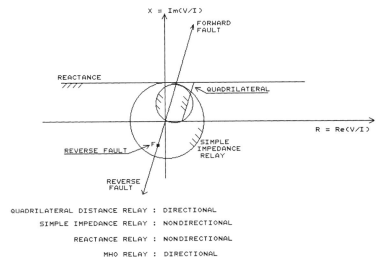

Figure 4.69 Directional feature.

4.7.2. Underreaching on Arcing Faults

Consider a line subjected to either a metallic fault or an arcing fault. Distance relays are set under the assumption that the fault is metallic. The question is, how will the relay behave for an arcing fault?

Figure 4.70 shows the SIR, reactance, MHO and QDR, all set to reach the same length of the line on a metallic fault. Assume an arcing fault, with arc resistance R_a, moving from the relay location to the reach point. We now have the fault characteristics of the line, or the impedance seen by the relay with and without R_a.

The amount of underreach for all relays is (a) Line length AB for SIR and MHO relay and (b) No underreach for reactance and QDR. It is also clear that the underreach is greater in MHO than in SIR. The following chart summarizes the situation:

Relay	Underreaching on arcing faults
SIR	Moderate
Reactance	None
MHO	Maximum
QDR	None

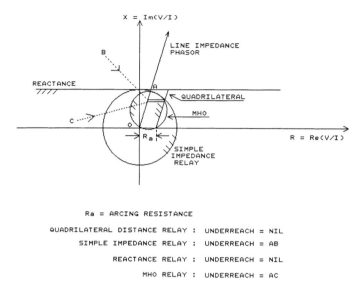

Figure 4.70 Underreaching on arcing faults.

4.7.3. Stability on Power Swing

Figure 4.71 shows the relay characteristics set to reach the same length of the transmission line. Assume that the line is now subjected to a power swing from which it can recover. We want to determine the maximum power angle δ_{max} which the various relays can accommodate without malfunctioning.

1. The reactance relay, unless directionalized by some additional directional relay, operates on virtually any value of δ. Thus, the reactance relay is the most unstable on power swing.
2. The SIR can accommodate some value of power angle δ_{max} as shown in Fig. 4.71.
3. The MHO relay can accommodate δ_{max} more than the SIR.
4. The QDR is the best relay, since it can accommodate a fairly large value of power swing.

This stability on power swing is shown in the following chart:

Relay	Stability on power swing
Reactance	Least
SIR	Moderate
MHO	More
Quadrilateral	Most

Chapter 4

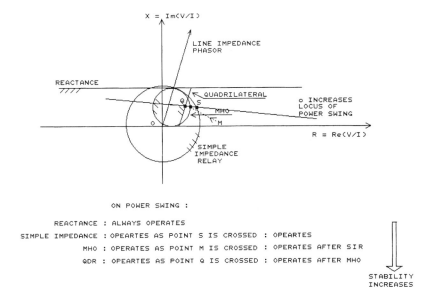

Figure 4.71 Stability on power swing.

4.8. THREE-STEPPED DISTANCE SCHEME

The purpose of any distance relaying scheme is to provide primary protection to the line under protection and backup protection to all adjoining power system elements. Refer to Fig. 4.72, which shows line AB under protection by relaying scheme R_A along with an adjoining line BC. The relay R_A should provide high-speed primary protection to line AB and back up the entire adjoining line BC. Line BC has its own distance protection.

4.8.1. Overreaching on Transient Faults and Setting Zone 1

Any distance relay (i.e., SIR, reactance, MHO, QDR, etc.) is prone to overreach on a transient fault consisting of a dc offset. As proved in Chapter 3, all high-speed relays tend to see more current because of the dc offset.

$$I_r = \sqrt{I_{ac}^2 + I_{dc}^2} \quad \text{with offset}$$

$$I_r = I_{ac} \quad \text{without offset}$$

The RMS value of fault current with a dc offset is greater than the RMS value of a pure alternating current. Therefore the impedance seen by the relay appears smaller (i.e., fictitious) than the actual value, as given by

Distance Protection

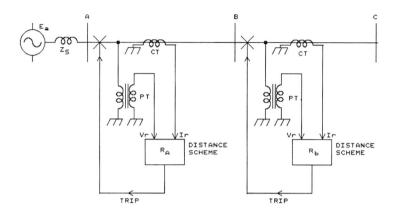

PURPOSE OF DISTANCE SCHEME R_a: PRIMARY PROTECTION OF AB AND BACK-UP FOR BC

Figure 4.72 Single line diagram/distance scheme.

$$Z_{r,\text{offset}} = \frac{V_r}{I_r} = \frac{V_r}{\sqrt{I_{ac}^2 + I_{dc}^2}}$$

$$Z_{r,\text{no offset}} = \frac{V_r}{I_r} = \frac{V_r}{I_{ac}}$$

$$Z_{r,\text{offset}} < Z_{r,\text{no offset}}$$

Refer to Fig. 4.73, where the distance relay is adjusted to reach point $X = 10.0$ ohms, under no-offset fault current. The source impedance is neglected. The fault current for a fault at X is

$$I_{r,\text{no offset}} = \frac{100 \text{ V}}{10 \text{ ohms}} = 10 \text{ A}$$

where $Z_n = 10$ ohms is the setting or steady-state reach of the relay.

Now move fault pont X to a new point $Y = 12.5$ ohms and let the fault current have a dc offset of 6.0 A. Thus,

$$I_{ac} = \frac{100 \text{ V}}{12.5 \text{ ohms}} = 8.0 \text{ A}$$

Since $I_{dc} = 6.0$ A,

$$I_{RMS} = \sqrt{8.0^2 + 6.0^2} = 10.0 \text{ A}$$

$$Z_{r,\text{fictitious}} = \frac{100 \text{ V}}{10.0 \text{ A}} = 10.0 \text{ ohms}$$

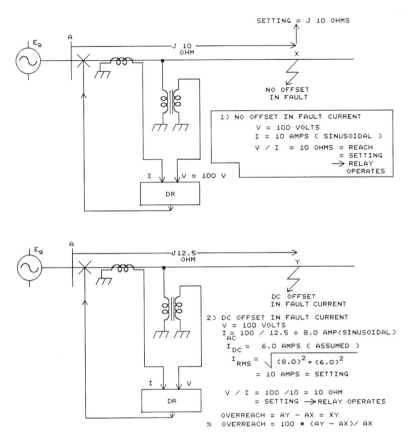

Figure 4.73 Transient overreach of distance relay. E_g = 100 volts.

The fictitious impedance due to the dc offset now equals the setting and the relay operates.

$$\text{Percentage overreach} = \frac{12.5 - 10.0}{10.0} \times 100 = 25\%$$

The better the design of the relay, the less is its transient overreach. Relay designers have not been able to reduce this overreach to less than 10% of its setting (i.e., steady-state reach). In practice, the percentage overreach is taken between 10% and 20%. Therefore, we cannot set the distance relays to 100% of the line length to provide high-speed primary protection to the entire line. If they are set to 100% and R_A overreaches on a transient

Distance Protection

fault, it will lose its selectivity w.r.t. relay R_B for faults on line BC. This is clearly shown in Fig. 4.74.

The high-speed distance measuring unit, which is now adjusted to 80% to 90% of the line length, is called the zone-1 relay. The margin of 10% to 20% is left for

Transient overreach
Calculated line impedance not equal to actual impedance
CT/PT phase angle and ratio error

We thus conclude that a Zone-1 relay is high-speed and must be adjusted to reach 80% to 90% of the protected line. It essentially provides primary protection.

4.8.2. Setting Zone 2

Since zone 1 has to reach 80% to 90% of the line length, the remaining 20% to 10% has no primary protection. Hence, an additional relay, called zone 2, is needed. A zone-2 relay needs to be delayed in time T_1 so that it is selective with zone 1, which is high speed, of the adjoining line.

1. The basic purpose of a zone-2 relay is to provide primary protection (delayed of course) to the remaining 10% to 20% of the line

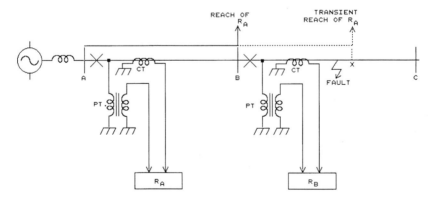

Figure 4.74 Loss of selectivity if first zone is adjusted to 100% of line length.

Chapter 4   157

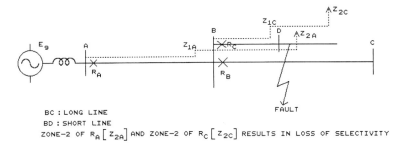

Figure 4.75 Incorrect setting of zone 2.

length and to provide partial backup protection to the adjoining line. The intentional time delay provided is around 0.3 to 0.5 sec.
2. The reach of zone 2 is taken as 25% to 50% of the adjoining shortest line, if there is more than one adjoining line.

This requires further explanation. Refer to Fig. 4.75. Line AB is under primary protection and BC and BD are the adjoining lines; BC is long compared to BD. Consider the wrong setting of zone 2 to be 50% of the longest line BC. Lines BC and BD have their own distance protection.

Note the encroachment of zone 2 of line AB into zone 2 of the shorter line BD. If a fault takes place in this overlapped zone, as shown at point F, line AB, as well as the shortest line BD, will be tripped out. This sort of loss of selectivity is absolutely undesirable. Hence, zone 2 should reach 25% to 50% of the *shortest* line. The correct setting is shown in Fig. 4.76.

4.8.3. Setting Zone 3

The zone-2 setting does give backup protection to adjoining lines to some extent. The distance scheme, however, requires additional zone 3 as backup,

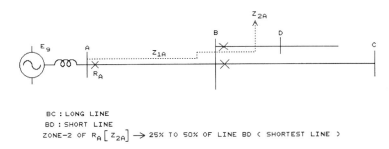

Figure 4.76 Correct zone-2 setting.

Distance Protection

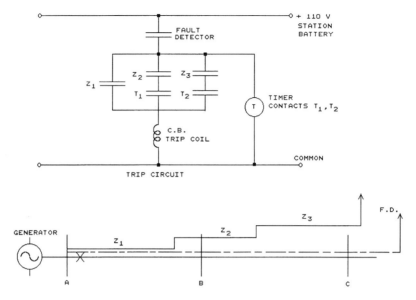

Figure 4.77 Three-stepped distance scheme.

with an additional delay of T_2, to all adjoining lines and power system elements. The rule is: The purpose of zone 3 is to provide 100% backup to all adjoining lines, and it must be discriminative in time with zone 2 of all adjoining lines. The intentional time delay provided is around 1.0 sec.

This reach of zone 3 is clear since all the lines need adequate backup. The overall settings of all zones are shown in Fig. 4.77, including the trip scheme.

4.9. LIMITATIONS OF THREE-STEPPED DISTANCE SCHEME

4.9.1. Faults on Entire Line Not Getting High-Speed Protection Requirement of Carrier

Figure 4.78 shows line AB with sources at both ends protected by the conventional three-stepped distance scheme by relays R_A and R_B. Note that faults in the middle 60% of the line (neglecting the underreaching of zone 1 on arcing faults) will be cleared instantaneously and simultaneously from both ends. This middle 60% will further decrease for arcing faults.

Duration of disturbance = (zone-1 time + breaker time)

Chapter 4

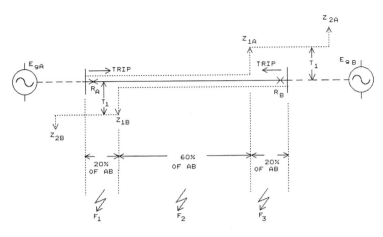

F_1 : FAULT CLEARED FROM END B BY DELAY OF T_1
F_2 : FAULT CLEARED FROM BOTH ENDS SIMULTANEOUSLY AND INSTANTANEOUSLY
F_3 : FAULT CLEARED FROM END A BY DELAY OF T_1

Figure 4.78 Three-stepped distance protection to double-end-feed lines.

Faults in the 20% at each end will be cleared instantaneously from one end but delayed by zone 2 from the other end. If I_A and I_B are the fault currents contributed from ends A and B, then the sequence for clearing the fault current is as follows:

1. $t = 0$: fault takes place in end 20% of line; total fault current $I_T = I_A + I_B$.
2. t = zone-1 time: fault cleared from end A instantaneously; fault current = I_B.
3. t = zone-2 time: fault cleared from end B with delay of T_1; fault current = zero.
4. Duration of disturbance = T_1 + CB time

Thus, faults in the end 20% create a power balance disturbance for a much longer time in order for automatic CB reclosure to be successful. What is desired is that Faults on the entire line be cleared instantaneously and simultaneously from both ends in order for automatic reclosure to be successful. This is achieved by using a carrier and is dealt with in another chapter.

4.9.2. Longest Line That Can Be Protected

The longest line that a distance relay can protect depends on the relay's performance during stable power swing. In other words, if the protected line is subjected to a stable power swing (i.e., it can recover by itself from the swing and maintain synchronization) and the distance relay malfunctions, the relay is incapable of accommodating maximum swing angle δ_{max}.

It is well known that a long line is less stable than a short line; that is, a long line has a larger swing angle δ_{max} compared to a short line. Figure 4.79 shows power-angle diagrams for long and short lines. The short line has higher P_{max} than that of a long line. For a given mechanical input P_m, the quiescent operating angle for a long line is more than that of a short line.

$$\delta_{o,long} > \delta_{o,short}$$

Therefore, for a disturbance that does not cause instability, the longer line goes through a larger swing angle than that does a short line. Hence, a long line is more unstable than a short line.

$$\delta_{max,long} > \delta_{max,short}$$

The longer the line, the greater is δ_{max}. The question is whether δ_{max}, on a stable power swing, can be accommodated by the relay without malfunctioning.

Let us now define *short line* and *long line*. In relaying parlance, a short or long line is not defined based on its length but rather on the system

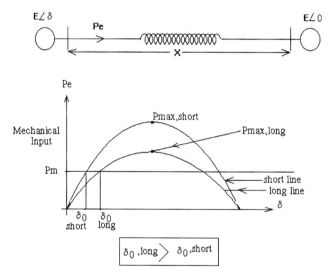

Figure 4.79 Quiescent operating angles for short and long lines.

impedance ratio (SIR) = Z_s/Z_l, where Z_s is the impedance behind the relay location and Z_l is the line impedance.

1. long line: Z_l is large, so the SIR = Z_s/Z_l is small
2. short line: Z_l is small, so the SIR = Z_s/Z_l is large

For brevity assume that

1. $Z_s = 0$ for a long line since SIR is small.
2. Z_s is finite for a short line since SIR is large.

Figure 4.80 shows these cases, with a MHO relay adjusted to 100% of line length (for simplicity the reach is assumed to be 100% rather than 80%).

It is clear that the δ_{max} that the MHO relay can accommodate without malfunctioning is around 90° for a long line, whereas it is greater than 90° for a short line. This is contrary to what we desire. A long line goes through a larger value of δ_{max}, and the relay must accommodate this. But the relay can accommodate only around 90°. Hence, There is a limitation to the longest line a relay can protect without malfunctioning on stable power swing.

4.9.2.1. Numerical Example

A transmission line is protected by an MHO relay, with zone 1 is adjusted to 100% of its length. The source impedances are $J10.0$ ohms. The line impedance is $J0.5$ ohm/mile. Find, in miles, the longest line that the relay can protect without malfunctioning on a stable power swing angle of $\delta_{max} = 120°$.

Answer: Refer to Fig. 4.81, which shows the angle line diagram of the network and the construction of the power swing locus. Examine triangle CPB.

$$\angle CPB = \frac{\delta_{max}}{2} = \frac{120°}{2} = 60°$$

$$\frac{BC}{CP} = \frac{BL + X_L/2}{X_L/2} = \tan 60° = 1.7321$$

$$\frac{1.7321 X_L}{2} = BL + \frac{X_L}{2} = J10.0 + \frac{X_L}{2}$$

$$X_L(1.7321 - 1.0) = J10.0 \times 2.0$$

$$X_L = \frac{J20.0}{0.7321} = J27.32 \text{ ohms}$$

$\therefore$ Longest line that can accommodate $120° = \dfrac{J27.32}{J0.5} = 55$ miles

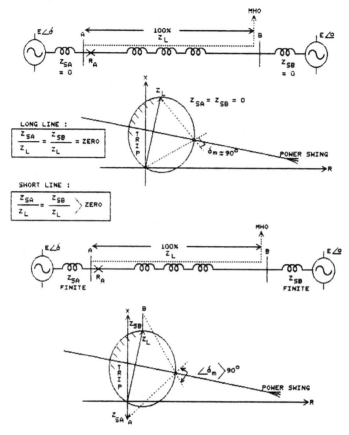

Figure 4.80 Power angle accommodation in short and long lines.

4.9.3. Shortest Line That Can Be Protected

Figure 4.82 shows a single line diagram with relay locations. Z_s is the source impedance. The distance relay is energized by V_r and I_r. Assume the distance relay is MHO set to reach Z_n along the line. If a fault takes place at the reach point, the relay voltage drops to a value depending on Z_s: the larger Z_s, the shorter is the line, by definition of system impedance ratio. Thus, the shorter the line, the smaller is V_r fed to the relay. An ideal distance relay is supposed to measure V/I, irrespective of individual magnitudes of either V or I. But in practice it is not so.

We shall prove that a MHO relay's reach is affected by the voltage V fed to it: the reach decreases as voltage decreases. Thus, on the faulted line

Chapter 4

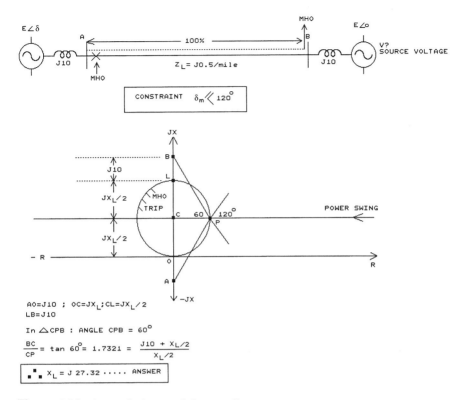

Figure 4.81 Numerical example/longest line.

due to voltage drop the actual reach will be less than the setting (i.e., theoretical reach). This decrease in reach is, essentially, due to a small restraining spring in the induction-cup-based MHO relay.

The theoretical trip equation is

$$\underbrace{|Z_n| \, |V| \, |I| \cos(\theta_n - \theta_r)}_{\text{operating torque}} > \underbrace{|V|^2}_{\text{restraining torque}}$$

Because of an additional restraining spring, there is an additional torque of T_s. The equation then becomes

$$|Z_n| \, |V| \, |I| \cos(\theta_n - \theta_r) > |V|^2 + T_s$$

Dividing both sides by $|V| \, |I|$ gives

$$|Z_n| \cos(\theta_n - \theta_r) > |Z| + \frac{T_s}{|V| \, |I|}$$

Distance Protection

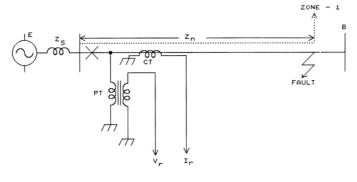

$$I_R = \frac{E}{Z_S + Z_n}$$

$$V_R = E - I_R * Z_S = E - \frac{E\, Z_S}{Z_S + Z_n}$$

$$V_R = \left[\frac{Z_n}{Z_S + Z_n}\right] E$$

$$V_R = \left[\frac{1}{1 + SIR}\right] E \quad ; \text{ where } SIR = \frac{Z_S}{Z_n}$$

DUE TO FAULT THE RELAY VOLTAGE DROPS FROM E TO V_r

Figure 4.82 Shortest line an MHO relay can protect.

Substituting $|I| = |V|/|Z|$ gives

$$|Z_n| \cos(\theta_n - \theta_r) > |Z| + \frac{T_s |Z|}{|V|^2}$$

To find the reach along the diameter, let $\theta_r = \theta_n$:

$$|Z_n| > |Z|\left(1 + \frac{T_s}{|V|^2}\right)$$

or

$$|Z| \leq \frac{|Z_n|}{1 + T_s/V^2}$$

The actual reach, considering restraining spring torque, is

$$\text{Actual reach} = \frac{Z_n}{1 + T_s/V^2} \quad \text{(function of } V\text{)}$$

Chapter 4

To check the answer, neglect the spring torque so that $T_s = 0$. Then

Actual reach = Z_n, which is constant and independent of V

Figure 4.83 shows the reach-versus-voltage characteristics of the MHO relay. Similar proofs exist for other relays.

The reach versus relay voltage is a slightly inconvenient characteristic, since the reach Z_n is required to be changed depending on the length of the line. To make the actual reach independent of $|Z_n|$, divide it by $|Z_n|$

$$\frac{\text{Actual reach}}{|Z_n|} = \text{Accuracy of distance relay} = x$$

To eliminate voltage, for a fault at the reach point Z_n, we calculate

$$V_r = E + IZ_s = E - \frac{E}{Z_s + Z_n} Z_s$$

$$= \frac{EZ_n}{Z_s + Z_n}$$

Dividing denominator and numerator by Z_n gives

$$V_r = \frac{1.0}{1.0 + Z_s/Z_n} E$$

$$= \frac{1.0}{1.0 + \text{SIR}} E$$

$$= \frac{1.0}{1.0 + y} E$$

where $Z_s/Z_n = y$ = system impedance ratio = range.

Thus, instead of plotting reach versus voltage, we plot accuracy versus range. Note that for small system impedance ratio or range or y, the relay voltage $V_r \rightarrow E$, the rated voltage. For larger system impedance ratio or range or y, then $V_r \ll E$.

Replotting the curve of Fig. 4.83 in terms of accuracy versus range, we obtain Fig. 4.84. The higher the source impedance or range, the lower is the actual reach or accuracy. Also the shorter the line, the higher is the SIR or range, thereby decreasing the relay's accuracy of reach. If one puts a limit that accuracy should not be less than 0.9 (i.e., reach not less than 0.9 times the actual setting), then there is a limit to the shortest line that a distance relay can protect.

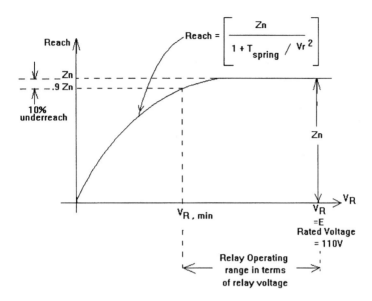

Figure 4.83 Reach-versus-relay voltage characteristics.

4.9.3.1. Numerical Example

Under no-fault conditions, a distance relay connected to a line has an accuracy of 1.0 for a rated voltage of 110 V to its pressure circuit. The line impedance is 0.5 ohm per mile. The source impedance is $J10.0$ ohms. Its accuracy falls to 0.9 (underreach by 10%) for a relay voltage of 10 V for a fault at the reach point Z_n. Calculate (a) the maximum SIR or range to which the relay can be applied and (b) the shortest line the distance relay can protect.

Answer:

$$V_{r,min} = \frac{1}{1 + y_{max}} E$$

$$10 = \frac{1}{1 + y_{max}} \times 110$$

Solving for y_{max}:

$$y_{max} = 10 = \frac{Z_s}{0.9 Z_1} \quad \text{(assuming reach = 90\% of } Z_1\text{)}$$

Therefore,

Chapter 4

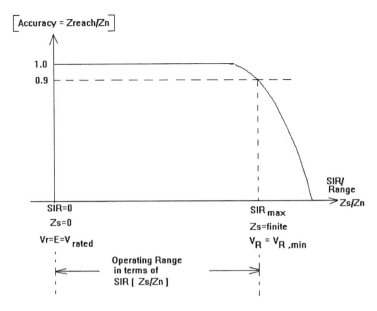

Figure 4.84 Accuracy-versus-range graph.

$$Z_1 = \frac{J10.0}{0.9} = 1.11 \text{ ohms}$$

and

$$\text{Shortest line} = \frac{1.11}{0.5} = 2.2 \text{ miles}$$

4.9.4. Difficulties in Protecting Double-End-Feed Lines

Most engineers had believed that the quadrilateral distance relay was probably the best solution for protecting HV and EHV lines. Unfortunately, the QDR also misbehaved and some of its malfunctions could not be explained at that point in time. Further studies revealed lacunas when it is used on DEF lines.

For DEF lines the fault characteristics of HV and EHV lines on the *R-X* diagram are found to swiveling QDR. The amount of swiveling and expansion depend on prefault export and import of power. Consider a single line diagram of a tie line between buses A and B. Let us confine our attention to relay R_A.

168   **Distance Protection**

The line is now protected by QDR R_A. We want to know how the QDR behaves on this DEF tie line. The fault characteristics of the line, under prefault and postfault export conditions, are superimposed on the R-X diagram.

4.9.4.1. Prefault Export/Import Conditions

Refer to Fig. 4.85. The fault characteristics of the line swivel in a clockwise direction, and their expansion depends on the arc resistance R_{arc}. Consider a fault beyond the reach point with R_A. The figure shows that the relay may now operate for certain values of R_A, causing the QDR to overreach.

Now consider an arcing fault immediately after the relay location. The impedance seen by the relay may get out of the tripping region, causing the relay not to operate. The terminal faults may not be cleared. Figure 4.85 shows how the operating characteristics shift to the right, showing clearly the overreach and portion of line, close to the relay location, not having the desired protection.

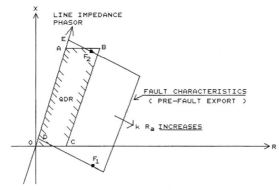

i) TERMINAL FAULT AT D NOT CLEARED (POINT F_1 IN FIGURE)
ii) AE OVERREACH (FAULT AT E , RELAY OPERATES ,(POINT F_2) IN FIGURE)

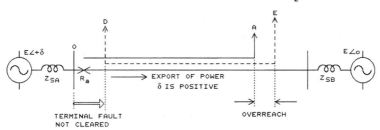

Figure 4.85 Malfunctions of QDR on prefault export conditions.

Chapter 4

Prefault Import Conditions

Figure 4.86 shows the QDR characteristics and the superimposed fault characteristics under prefault import conditions. The fault characteristics now swivel in counterclockwise. Consider an arcing fault below the reach point, as shown in the figure. Some value of R_A will take the impedance seen by the relay outside the trip region of the QDR, causing the relay to underreach.

Consider another fault, this time behind the relay location as shown. The value of R_A may push the relay impedance inside the tripping region, causing the QDR to lose its directional feature.

In effect, it is possible for a QDR to underreach and lose its directional feature under prefault import conditions. The relay characteristics shift to its left, showing clearly the underreach and loss of directional feature.

The natural solution is to have a swiveling QDR, whose swiveling and expansion depend on prefault power flow conditions. The solution is quite difficult and falls in the realm of adaptive relaying.

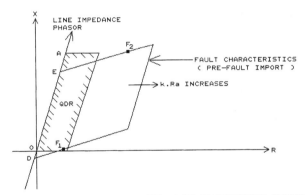

i) REVERSE FAULT AT D : RELAY OPERATES : LOSS OF DIRECTIONAL FEATURE
ii) FAULT AT E : RELAY DOES NOT OPERATE (POINT F_2) : UNDERREACH

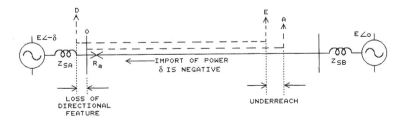

Figure 4.86 Malfunctions of QDR on prefault import conditions.

4.10. OUT-OF-STEP BLOCKING AND TRIPPING SCHEMES

As pointed out earlier, in a two-machine problem there are two types of power swings. The first one is due to a small external disturbance outside the tie line under protection, causing a stable power swing from which the two machines can recover and remain synchronous. If the distance relay malfunctions on such a stable power swing, the system is in trouble, resulting in total dislocation. In this situation an out-of-step blocking scheme is needed.

For a large external disturbance the power swing is unstable, and the two machines must be separated. The operation or nonoperation of distance schemes, in a superficial way, appears to be of no consequence. But this is not true. The point of ultimate split of the two machines should not be left to haphazard operation of distance relays, but the point of split must be decided in advance so that the separated systems have minimum load dislocation. At the predetermined split point, we need a deliberate out-of-step tripping scheme. The following example, shown in Fig. 4.87, clearly indicates the locations where out-of-step blocking or tripping schemes are needed. Two generators, G_A and G_B, are connected over a tie line AC with an intermediate bus B. The loads are

bus A, L_A = 200 MW
bus B, L_B = 300 MW   total load = 700 MW
bus C, L_C = 200 MW

The tie-line power flows are

Bus A to bus B = 200 MW

Bus C to bus B = 100 MW

The power generation is

Generator G_A = 400 MW

Generator G_B = 300 MW

Total generation = 700 MW

The generator and line reactances are

Generator G_A = $J0.2$ pu

Generator G_B = $J0.2$ pu

Line AB = $J0.4$ pu

Line BC = $J0.2$ pu

Chapter 4

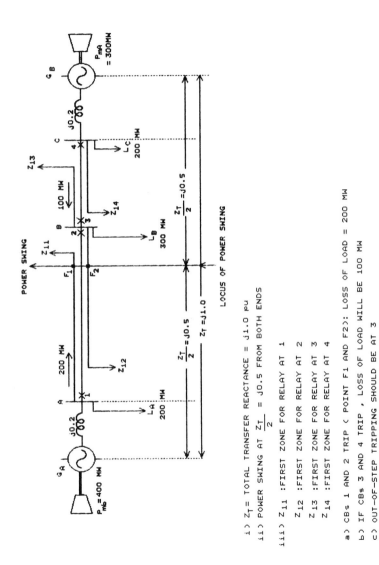

Figure 4.87 Locations of out-of-step blocking and tripping schemes.

In the two-machine problem the power swing passes through the center of the system (i.e., $Z_t/2$); that is, it is equidistant, in terms of total transfer reactance, from both sources. Since total impedance is $J1.0$ pu, the power swing passes through $J0.5$ pu from both sources.

Let us now look at the haphazard operation of the four distance schemes (at 1, 2, 3 and 4) and the dislocation it causes to the loads. We consider the operation or nonoperation of zone-1 units alone of distance relays at 1, 2, 3 and 4. The relay characteristics may be assumed to be of MHO type, and the zone-1 reaches are shown in the figure.

The power swing passes through zone 1 of R_A and R_B, causing the line breakers to trip. The power swing is behind zone 1 of the relay at 3 and outside zone 1 of the relay at 4. Thus, no relays nor associated CBs operate on the power swing. The situation is that CBs 1 and 2 are tripped, causing the system to split into two subsystems, thereby resulting in loss of tie-line power of 200 MW over line AB.

The correct solution is to choose the split point at either 3 or 4 or both so that the two split systems have, as far as possible, their generation and load more or less matched. This will ensure minimum dislocation to the load. Thus, if the system is split at point 3, the power lost is 100 MW, due to loss of line BC. In effect, then, On a power swing the system should not be left to itself to split, by haphazard relay operation, thus causing wide dislocation to the loads. The split point should be predetermined on the basis of power flows, so that the load dislocation is a minimum. Having decided the split point, incorporate an out-of-step tripping relay at that point only, and at all other relay locations use out-of-step blocking relays.

4.10.1. Out-of-Step Blocking Scheme

As pointed out earlier, an out-of-step blocking relay is required where the zone-1 distance relay is likely to operate, but we wish to block the distance scheme at that location so that the system is separated elsewhere at the correct point.

Further, the location of the locus of power swing depends on source and line impedances (i.e., perpendicular bisector of $Z_t = Z_{sA} + Z_1 + Z_{sB}$). Although the line impedance is constant, the source impedances are likely to vary from time to time. This is due to either dropping or increasing generating units or line tripping behind the relay location. Thus, the location of power swing cannot be precisely predetermined.

The power swing is assumed to be in the trip region of zone 1, thereby causing the relay to trip and possibly split the system at the wrong place. To avoid this an out-of-step blocking relay is used (Fig. 4.88). The out-of-step blocking relay has circular characteristics and is concentric to zone 1.

Chapter 4

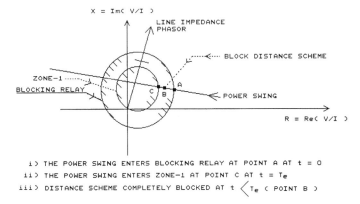

Figure 4.88 Out-of-step blocking relay.

Many engineers also believe that zone 2 also may operate on the swing, in which case the out-of-step relay has to be concentric to zone 2 rather than zone 1. For our discussion, we take it as concentric to zone 1. This ensures that the power swing will always enter the blocking relay before it enters zone 1.

The operation is now quite simple. Let the power swing enter the blocking relay at $t = 0$. At $t = T_e$ it enters zone 1. The whole distance scheme is blocked or made nonoperational by the blocking scheme at time less than

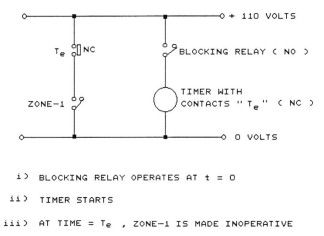

Figure 4.89 Trip scheme with out-of-step blocking relay.

T_e, or $t_{blocking} < T_e$. The trip circuit of the three-stepped-distance scheme, along with the contacts of the blocking relay, are shown in Fig. 4.89.

Since the blocking relay prevents the relay scheme from operating, it is desirable to prove that the blocking relay does not interfere in the relay scheme for genuine internal faults on the line. Consider an internal line fault, not a power swing. The power swing is basically a slow phenomenon, due to inertia of machines, and therefore it enters the blocking relay first; later (after time T_e) it enters zone 1.

During a fault, the impedance seen by the relay abruptly changes from Z_{load} to Z_{fault}. Before the blocking relay makes the relay nonoperational, the relay scheme operates and genuine internal faults are cleared.

4.10.2. Out-of-Step Tripping Scheme

When the generators are not synchronous, all ties between them should be opened to maintain service and to permit the generators to be synchronized. The separation should be made only at such locations that the generating capacity and the loads on either side of the point of separation will be evenly matched so that there will be minimal interruption to the service. This deliberate tripping does not depend on whether the distance scheme operates. For this reason the power swing, whose location is variable, is assumed to be outside zone 1, as shown in Fig. 4.90.

The out-of-step tripping scheme, therefore, consists of two blinder relays B_1 and B_2, having straight-line characteristics and parallel to the line impedance phasor (Fig. 4.91). The hatched region is the tripping region.

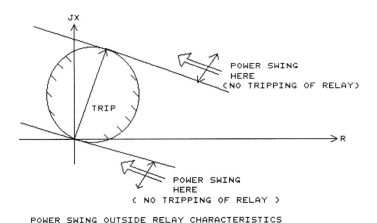

Figure 4.90 Location of power swing for tripping scheme.

Chapter 4

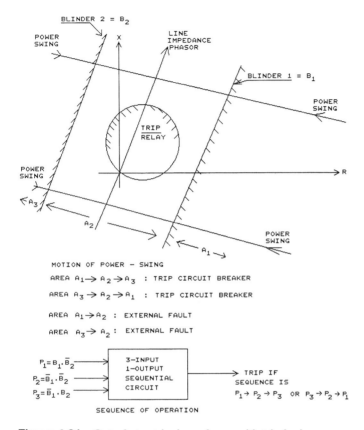

Figure 4.91 Out-of-step tripping scheme with trip logic.

The blinders split the R-X diagram into three distinct areas A_1, A_2 and A_3. In case of a power swing, the locus enters the three areas in the following way:

		—time increases →		
Power swing from right to left	→	A_1	→ A_2	→ A_3
Status of blinder contacts	B_1 →	closed	open	open
	B_2 →	closed	closed	open
Power swing from right to left	→	A_3	→ A_2	→ A_1
Status of blinder contacts	B_1 →	open	open	closed
	B_2 →	open	closed	closed

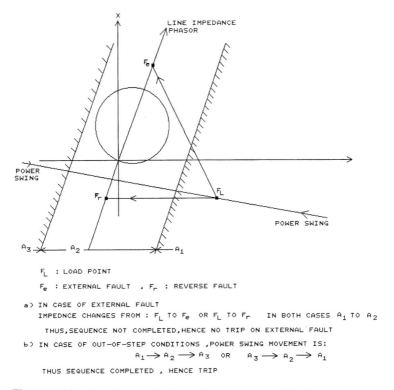

Figure 4.92 Out-of-step tripping scheme on external fault.

Thus, if the time sequence of blinders is one of the foregoing, the trip circuit overrides the relay trip circuit and trips the circuit breaker. Figure 4.91 shows the trip circuit.

Since this tripping scheme has a definite purpose, it should not trip the CB on external faults. Figure 4.92 shows an external fault at F_e and a reverse fault at F_r behind the relay location. It is clear that sudden changes in impedances, because of a fault, do not cause both the blinders to operate. Therefore, the out-of-step tripping scheme is made ineffective, and the normal distance scheme takes over. No CBs trip on either external or reverse faults.

4.10.3. Numerical Example

Refer to Fig. 4.93, which shows a tie line AC between sources E_A and E_B, with an intermediate bus B. CBs, various impedances and loads. (i) If no

Chapter 4

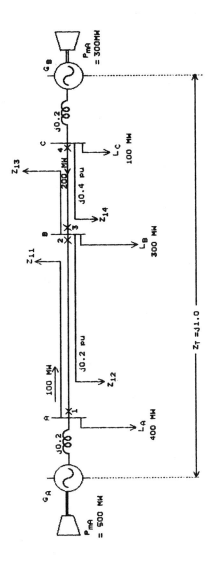

Figure 4.93 Numerical example.

additional out-of-step blocking and tripping schemes are used, which CBs operate? Assume MHO relays and power swing in zone 1 only. What is the magnitude of load dislocation? (ii) Where will the out-of-step tripping scheme and out-of-step blocking scheme be located?

Answer:
    (i) CBs 3 and 4 will be tripped, with dislocation of tie line BC resulting in interruption of 200-MW load on the intermediate bus.
    (ii) The point of separation should be at CB1, and thus it is the location of the out-of-step tripping scheme. All other locations should have an out-of-step blocking scheme.

4.11. PROBLEMS AND EXERCISES

1. Explain why the combination of instantaneous and DTOC cannot be used for HV and EHV lines.
2. Derive the fault characteristics of HV and EHV lines to accommodate metallic and arcing faults for (i) single-end-feed line, (ii) double end feed with prefault export of power, (iii) double end feed with prefault import of power.
3. From a simple impedance relay, reactance relay, MHO relay, and quadrilateral relay choose the best relay characteristics. Consider directional features, underreaching on an arcing fault, stability on power swing, etc.
4. In a three-stepped distance scheme why is
   a. Zone 1 adjusted to less than 100% of line under protection?
   b. Zone 2 adjusted to reach 25% to 50% of the shortest adjoining line?
   c. Zone 3 adjusted to reach beyond (about 110%) the longest line?
5. Using an induction cup, derive the torque relationship and place suitable polarizing, restraining and operating coils. What is the significance of polarization?
6. Derive the diameter or setting (i.e., Z_n) of the MHO circle as a function of the CT/PT ratio and difference in line impedance angle (i.e., θ_l) from the MTA θ_n of the MHO relay.
7. Derive the duality relationship between the cosine-type phase comparator and an amplitude comparator. Verify that it can be used to generate an MHO relay.
8. What is the coincidence period (positive or negative) for more than two sinusoidal inputs. How is this property used to develop a quadrilateral distance relay?
9. What other techniques can be used to generate a QDR?

Chapter 4

10. What are the advantages and disadvantages of static relays over electromechanical relays?
11. Why do we say that processor-based relays can be adaptive or generate many characteristics? Explain using the example of OC relaying.
12. Explain the following techniques for calculating impedance, resistance and reactance from voltage/current inputs. How do they perform on harmonic response and computational time?
    a. Fourier
    b. one-term Walsh
    c. two-term Walsh
13. How is the sampling rate tied to the processing time for processor-based distance relays? Why is it done?
14. How many measuring units are required to protect a three-phase line for all 10 possible short circuits?
15. Derive the inputs to a B-C phase-fault measuring unit so that it correctly measures the positive sequence impedance up to the B-C fault point. How does this measuring unit behave during B-C-G and three-phase faults?
16. Derive the inputs to an A-E ground-fault measuring unit, so at it correctly measures the positive sequence impedance up to the A-E fault point. How does it behave during a three-phase fault?
17. What is the difference between nonswitched and switched three-stepped distance schemes? What should be the reach of a fault detector? State which scheme is slower and give your reasons.
18. Sketch the diagram to trip the CB for a three-stepped distance scheme.
19. What are the limitations of distance relays from the following viewpoints:
    a. Speed of relaying for faults on the entire line
    b. Shortest line a distance relay can protect, in view of voltage drop to the relay due to a fault
    c. Longest line a distance relay can protect without malfunctioning on a stable power swing
20. How does the simple QDR, suitable for a single-end-feed line, behave on a double-end-feed line?
21. Using a suitable example of a two-machine system with a tie line and intermediate bus, locate the out-of-step tripping and blocking schemes.
22. It is said that (a) one has few out-of-step tripping relays (b) large numbers of out-of-step blocking schemes. Explain the reasons for this statement.

5
Carrier Schemes for HV and EHV Lines

5.1. INTRODUCTION

A fault analysis of overhead lines yields the following statistical data:

Transient faults 80%
Semipermanent faults 10%
Permanent faults 10%

Recall that a *transient fault* is one such as an insulator flashover, which is cleared by the immediate tripping of the circuit breaker and does not recur when the circuit breaker is reclosed, with predetermined delay. A *semipermanent fault* is one such as that caused by a tree branch falling on the line. Here the cause of the fault cannot be removed by the immediate tripping of the circuit breaker, but could be burnt away by repeated closing of the circuit breaker. Therefore the obvious advantage of automatically reclosing the circuit breaker is a reduction in the period of supply interruption. A further benefit, applicable to HV/EHV systems, is that system stability can be safeguarded.

The problems of applying autoreclosure schemes may therefore be conveniently discussed under two headings:

1. MV autoreclosure, where continuity of supply is the principal aim
2. HV autoreclosure, where problems of stability and synchronizing are of paramount importance

5.2. AUTORECLOSURE

5.2.1. Medium-Voltage Autoreclosure

The following are the benefits of autoreclosure in MV networks.

1. Supply interruptions to the consumer are minimized. The interruption duration is only a few seconds. This is true only for transient faults. Statistically, however, most faults are transient.
2. In unattended substations, the cost of personnel and time required for manual reclosure are eliminated.

Medium-voltage lines use *multishot* automatic circuit breaker reclosure. The *three-shot* reclosure is the most widely used, and its timing is as follows:

1. The first reclosure is instantaneous. About 80% of transient faults are normally cleared.
2. The second reclosure is made after a time delay, adjustable from 15 to 45 sec, and about 10% of the remaining transient faults are cleared.
3. The third closure is made after a time delay that is adjustable from 60 to 120 sec to clear 2% of the transient faults.

After the third reclosure, and if the fault is not cleared or burned out, the reclosing relay locks out. The auto reclosing scheme essentially uses a timing device such as a small synchronous motor (electromechanical device) to time the three reclosures.

5.2.2. High-Voltage Autoreclosure

All the schemes used on bulk transmission lines are high-speed, single-shot; that is, only one attempt is made to clear the fault and maintain the line in service. The reason is that the fault levels associated with these lines are extremely high and, apart from the arduous duty imposed on the circuit breakers, the shock to the system and danger to the generators drifting apart, leading to instability, are progressively increased by multishot schemes.

The successful application of high-speed single-shot autoreclosure to a high-voltage tie line, interlinking two synchronous systems, depends on four factors:

1. The maximum time permissible by the system for the opening or closing of the circuit breakers at each end of the tie line before the two systems drift apart leading to instability.
2. The time that must be allowed for the arc path to deionize so that it will not restrike when the circuit breakers are reclosed on the

faulty tie line. Deionizing times are also called dead times in the literature. Typical deionizing times for arcs for various voltage transmission lines are

Transmission line voltage (kV)	Deionizing time (sec)
66	0.10
132	0.17
220	0.28
400	0.50

3. The operating speed of circuit breakers while opening and closing.
4. The probability of transient faults that will allow high-speed autoreclosure of the faulty line without restriking of the fault.

In practice, a compromise is usually adopted between these conflicting requirements so that stability is not lost nor is outage time to the consumer excessively large. It is true that an unsuccessful reclosure is more detrimental to the system than no reclosure at all. For this reason the time to deionize the arc path has to be adequate to achieve the best results.

Relaying at both the ends must be high speed. This is not possible with conventional three-stepped distance schemes. A carrier is needed so that faults from both ends are cleared instantaneously and simultaneously. With high-speed relaying and high-speed circuit breakers, like air blast, the fault current is interrupted from both ends, in about 3 or less cycles of power frequency (i.e., 50 or 60 Hz).

The reclosing time is generally defined as the time taken by the circuit breaker to open and reclose the line and is measured from the instant that the relay energizes the trip coil of the breaker to the instant the breaker contacts remake the tie-line circuit. Reclosing time comprises breaker time and deionizing time (also called dead time).

Figure 5.1 shows the general sequence of operations for

1. Successful reclosure

    transient fault → high-speed trip → high-speed reclosure

2. Unsuccessful reclosure

    permanent fault → high-speed trip → high-speed reclosure → high-speed trip → lockout

184   **Carrier Schemes**

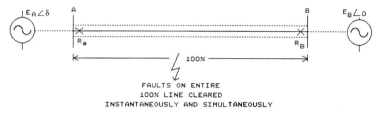

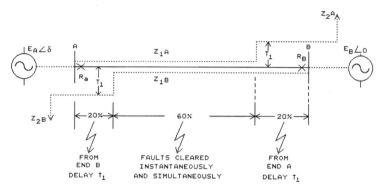

Figure 5.1 Successful and unsuccessful reclosures.

5.3. CONDITIONS ON RELAYING AND CIRCUIT BREAKER FOR HIGH-SPEED AUTORECLOSURE

Figure 5.2 shows the typical three-stepped characteristics of distance relays.

Observe that there is a 10% to 20% zone near each end of the line, in which faults are cleared by sequential tripping. These *end zones* usually represent about 20% to 40% of the line under protection. The remaining 60% to 80% (called the *middle* zone) between end zones is cleared simultaneously by the circuit breakers at both ends.

The delayed opening of CB from one end for faults in the end 10% to 20% of line invariably results in instability, due to delayed reclosure equal to operating time of zone 2. To overcome this difficulty—in other words, to ensure instantaneous and simultaneous opening of circuit breakers for faults over the entire line, one needs a carrier channel.

Chapter 5

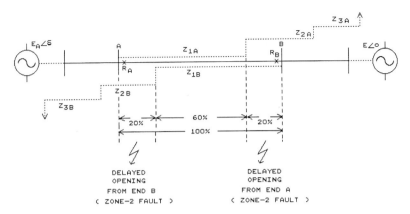

Figure 5.2 Three-stepped characteristics of distance relays.

5.4. CARRIER COUPLING

5.4.1. Introduction

Carrier coupling permits simultaneous and instantaneous CB tripping at both the ends of the tie line for faults on the entire line, thereby permitting high-speed autoreclosure. The carrier frequency is 50 to 500 kHz, well above the highest audio frequency and well below the lowest radio propagation frequency, thus ensures minimum interference.

5.4.2. Carrier Coupling Configurations

Figure 5.3 shows injection of a carrier into a single-phase conductor. It is far more common to use two phase conductors, giving phase-to-phase transmission of the carrier. Coupling ensures that the carrier is received in adequate strength.

High-frequency barriers, in the form of parallel resonant to chosen carrier frequency, called *line traps*, are installed at both ends of the line so that the injected carrier stays within the line and does not propagate to adjacent lines. This trap has negligible impedance to 50–60 Hz and is thermally designed to carry the power frequency load current.

Carrier injection on the power line and extraction from it are achieved through high-voltage capacitors used in conjunction with a drainage coil, to isolate terminal equipment from the high-voltage line.

The carrier injection equipment, known as a line filter, is completed by the addition of spark gaps to protect the components from line surges, an electric switch to permit maintenance on the filter, and means for impe-

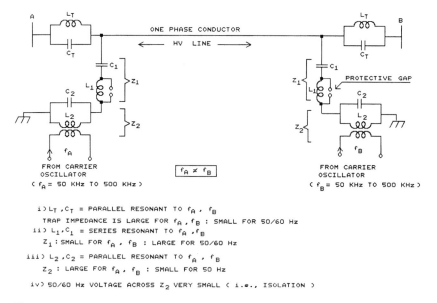

Figure 5.3 Single-phase carrier coupling.

dance matching of the high-frequency cable connecting the indoor equipment and the line filter. Figure 5.4 shows a complete phase-to-phase line coupling of the carrier coupling arrangement. For relaying purposes the carrier is not amplitude modulated but used only in on/off or injected/not injected modes.

5.5. CARRIER-AIDED DISTANCE SCHEMES

5.5.1. Introduction

Carrier signaling is used in connection with distance relays to speed up fault clearance, falling in delayed zone 2, from both ends. The delay inherent in the second zone may be overcome by transmitting a trip instruction to the relay at the remote terminal via a carrier channel, transmission being initiated by local zone-1 relay operation. The carrier facility thereby allows simultaneous and fast tripping at both ends of the tie line and permits application of autoreclosing schemes.

5.5.2. Transfer Tripping or Intertripping

The simplest way of speeding up the fault clearance time for a fault in zone 2 is to adopt a straight transfer trip or intertrip technique, where zone-1

Chapter 5

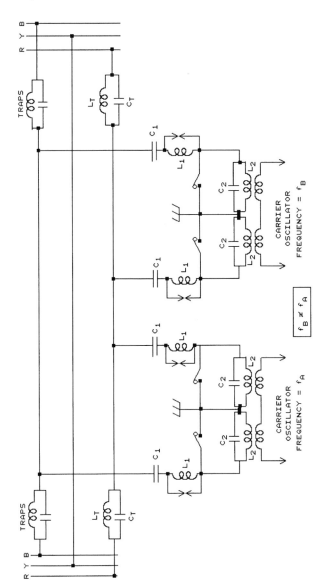

i) L_T, C_T PARALLEL RESONANT TUNED TO CARRIER

ii) L_1, C_1 SERIES RESONANT TUNED TO CARRIER

iii) L_2, C_2 PARALLEL RESONANT TUNED TO CARRIER

Figure 5.4 Phase-to-phase carrier coupling.

contact of the remote end is used to inject a carrier, and a carrier receipt relay A at the near end is used to initiate CB tripping.

Figure 5.5 shows the conventional three-stepped distance characteristics, a trip circuit at end A, and the carrier intertrip scheme. Consider a fault at X, which falls in zone 2 of local end A and in zone 1 of remote end B. The zone-1 relay contacts, while tripping its own circuit breaker at end B, also initiates or injects the carrier into the line. The carrier is received at end A by carrier receipt relay A (CRR_a) which shunts the series combination of fault detector, zone-2 relay contacts and zone-2 timer contacts (i.e., T_1) to initiate faster tripping. With this scheme the carrier is required to be transmitted over a faulty line, and therefore adequate carrier power needs to be injected.

Further, during external fault conditions, the carrier receipt relay may operate due to excessive noise generated within the protected zone and false

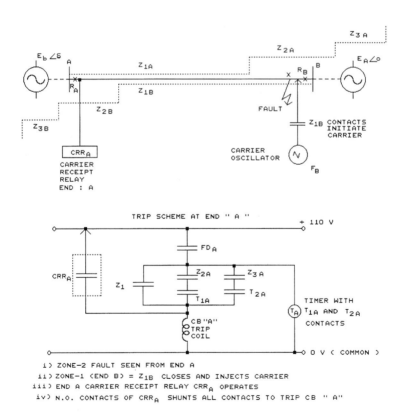

Figure 5.5 Transfer tripping or intertripping.

Chapter 5

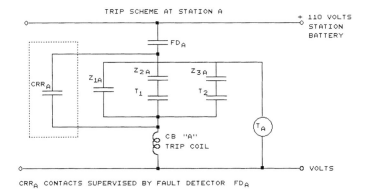

Figure 5.6 Permissive intertrip trip.

tripping may occur. To reduce false trips the CRR_a circuit can be monitored by the fault detector, as shown in Fig. 5.6, so that CRR_a performs its intended function during a genuine internal fault. This arrangement is called *permissive intertrip* scheme.

5.5.3. Carrier Acceleration

Refer to Fig. 5.7. The faults under consideration are zone-2 faults. A method of accelerating this zone is to shunt the zone-2 timer contacts by a normally open CRR_a in the end A trip circuit. Thus, all faults within the entire protected line can be cleared in approximately zone-1 time. This arrangement, known as *carrier acceleration scheme*, also depends on the capability to transmit the carrier over a faulted line.

5.5.4. Carrier Blocking or Preacceleration of Zone 2

The schemes described earlier have used the carrier to transmit a tripping instruction, and they revert to a conventional three-stepped-distance scheme in the event of carrier channel and equipment failure. A different arrangement, known as *carrier blocking*, is shown in Fig. 5.8. Consider a fault X at the end of line AB, which falls in zone 2 from end A. In the trip scheme at end A the normally closed contacts of CRR_a bypass the zone-2 timer contacts T_1, thereby preaccelerating zone 2 to virtually instantaneous operation. Therefore, faults on the entire line are cleared instantaneously. The carrier blocking scheme is therefore called preacceleration.

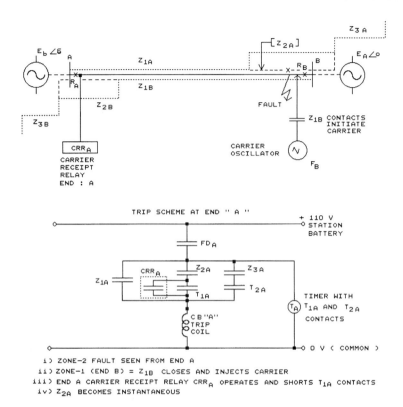

Figure 5.7 Carrier acceleration.

Since zone 2 also extends to the adjoining line, the carrier is required to be transmitted, for zone-2 faults external to the protected line, to prevent accelerated tripping.

At end B we have, in addition, a reverse relay RR_b, whose reach is adjusted beyond zone 2 of end A (Fig. 5.8). The normally open contacts of RR_b are used to inject a carrier so that CRR_a operates in the event of zone-2 faults, say at Y, in the adjoining line, and the NC closed contacts of CRR_a, shunting the timer contacts T_1, open. Zone 2 of the distance scheme at end A will revert to its normal operation (i.e., delay T_1). The carrier, therefore, is used to block the operation of zone 2 instantaneously for faults outside the line.

Note that the carrier is required to be transmitted over a healthy line. This is the greatest advantage. The disadvantage is that carrier failure results in loss of selectivity. Thus, maintenance is of paramount importance. Note

Chapter 5

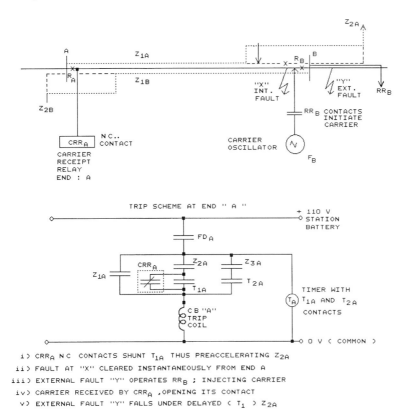

Figure 5.8 Carrier blocking/preacceleration of zone 2.

that all forward-looking relays need to be slightly delayed in time so that all reverse-looking relays have the first opportunity to detect the external fault.

5.6. UNIT CARRIER SCHEMES

Unit carrier schemes provide unit protection without backup protection. Unit relay schemes essentially compare the line conditions at the two ends and provide tripping, with the help of a carrier, in case the two conditions have drastically changed, indicating an internal fault on the line.

192 Carrier Schemes

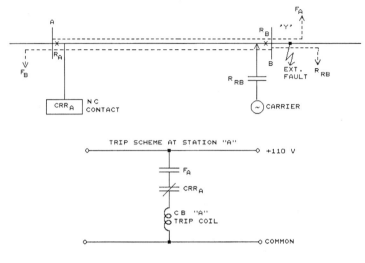

i) FAULTS ON ENTIRE LINE CLEARED BY F_A AND F_B
ii) EXT. FAULT AT Y OPERATES R_{RB} , INJECTING CARRIER
iii) CARRIER RECEIVED BY CRR_A ,ITS CONTACTS OPEN , THEREBY THE TRIP SCHEME IS INOPERATIVE

Figure 5.9 Basic power directional unit scheme.

5.6.1. Power Directional Unit Carrier Scheme

Figure 5.9 shows the basic principle, in which at each end of line there is a forward-looking and a reverse-looking relay. If the fault lies within the line, the forward-looking relays at end A (F_A) and end B (F_B) initiate simultaneous and fast tripping from both ends.

Now consider a fault at Y, which is an external fault on the adjoining line, beyond bus B. One has to ensure that the reach of the reverse-looking relay at end B (relay R_{RB}) is beyond the forward-looking relay at end A (relay R_{FA}). For an external fault at X, R_{RB} operates and initiates the carrier and deenergizes tripping at end A. The tripping arrangement at station A is also shown in the figure. It is clear that a slight delay must be provided to all forward-looking relays so that when the carrier is received CRR_a will have enough time to operate and prevent tripping.

Note that carrier failure causes complete loss of selectivity. Therefore, great attention must be given to the carrier equipment maintenance.

Chapter 5

5.6.2. Phase Comparison Scheme

The basic principle is to check the phase difference between the current at end A and that at end B, with the help of the carrier. A carrier is necessary because the ends of the line are miles apart. Refer to Figs. 5.10a, b and c, where the phase shift between the end A CT secondary current I_{AR} and the end B current I_{BR} are shown for

1. Normal power flow
2. External fault
3. Internal fault

Note carefully the current transformer polarity marks.

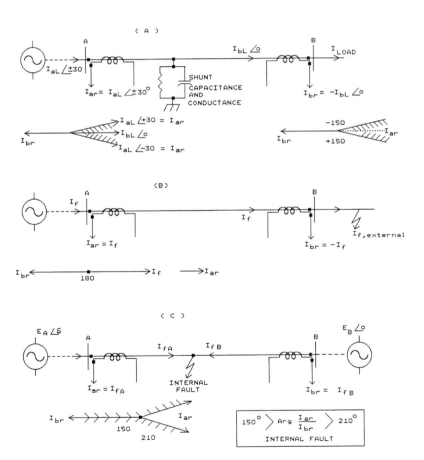

Figure 5.10 Phase shift between line and circuit: (A) normal power flow; (B) external fault; (C) internal fault.

It appears that under normal power flow conditions as well as external fault, the input line currently I_{AL} at end A and the output line current I_{BL} at end B should, more or less, be in phase. This is true on the line side and not on the CT secondary side, due to the chosen CT polarity marks.

Under normal load flow conditions, i_{AL} either leads or lags I_{BL} by around 0° to 30° due to shunt capacitances and line conductance. On the relay side the currents are designated I_{AR} and I_{BR}. Now the phase difference between I_{AR} and I_{BR} is shown in Fig. 5.11 for

1. Normal power flow

$$-150° > \arg \frac{I_{AR}}{I_{BR}} > +150°$$

2. External fault

$$\arg \frac{I_{AR}}{I_{BR}} = +180°$$

3. Internal fault

$$+150° > \arg \frac{I_{AR}}{I_{BR}} > +210°$$

The tripping region is shown hatched in Fig. 5.11, and the trip criterion in analytical form is

$$+150° > \arg \frac{I_{AR}}{I_{BR}} > -150° \text{ or } +210°$$

The question is how to compare the phase shift between the local current (say at end A) and the remote current at end B over a carrier and then take a trip or no-trip decision according to the trip criterion. The basic principle of checking the phase shift is based on the coincidence principle, as dis-

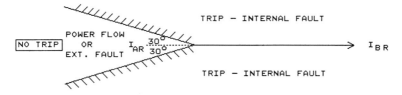

Figure 5.11 Phase shift between input/out line currents for various conditions.

Chapter 5

cussed in phase comparator theory in Chapter 3. The coincidence period between I_{AR} and I_{BR} is shown in Fig. 5.12.

The basic principle of comparing the phase shift is as shown in Fig. 5.13 and is described below.

1. Square the sinusoidal waveform of local current I_{AR} at end A.
2. Square the sinusoidal waveform of remote and current I_{BR} at end B and inject the carrier during the positive half-cycle of the square wave into the line. No carrier is to be injected during the negative half-cycle.
3. The carrier is received at end A and demodulated to recover the square wave, which is representative of the phase position of I_{BR}.
4. The local end square wave (represents the phase of I_{AR}) and the demodulated square wave (represents the phase of I_{BR}) are inputted to the logical AND gate.
5. If the pulsewidth of this output pulse from the AND gate is equal to or more than 1/12 cycle (corresponding to 30° or 1.65 msec on a 50-Hz base, or 1.4 msec on 60-Hz base), a trip is issued to trip the circuit breaker at A.
6. Since this scheme uses only one carrier, it is not possible to send three carriers, representative of the phases of the three line cur-

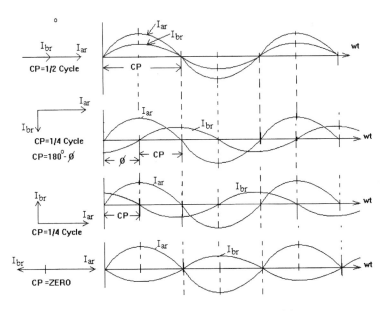

Figure 5.12 Coincidence period under various conditions.

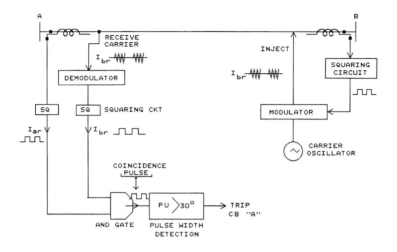

i) FIGURE SHOWS SCHEME AT BUS A
ii) SIMILAR SCHEME AT BUS B (NOT SHOWN)

Figure 5.13 Hardware of phase comparison scheme.

rents. For the scheme to work satisfactorily for all 10 faults, a single sinusoidal quantity, which is a function of all the line currents, is derived with the help of a summation transformer. This is shown in Fig. 5.14.

7. The general relation of the output current, which represents a single-phase quantity representative of three-phase currents, is

$$I_{\text{output}} = (N + 2)I_A + (N + 1)I_B + NI_C$$

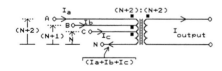

i) I_{output} = (N+2) Ia + (N+1) Ib + (N) Ic

 = $(a^2 + 2) I_{a1}$ + $(a+2) I_{a2}$ + 3 (N+1) I_o

ii) OUTPUT CURRENT IS A REPRESENTATIVE OF ALL THREE LINE CURRENTS
iii) IF I_{output} = ZERO : IT IS CALLED "BLIND SPOT"

Figure 5.14 Summation current transformer.

In terms of symmetrical components it is

$$I_{output} = I_1(2 + a^2) + I_2(2 + a) + I_0[3(N + 1)]$$

The choice of N is particularly controversial. It depends on no blind spots (i.e., I_{output} not zero) for all 10 possible faults on the given system. The other type of output sometimes employed is

$$I_{output} = 5I_2 - I_1$$

5.7. PROBLEMS AND EXERCISES

1. What limitations of three-stepped distance schemes require a carrier?
2. Describe carrier acceleration as used in a carrier acceleration scheme.
3. What is carrier preacceleration and in which scheme is it used?
4. In a power directional unit carrier scheme, why do forward-looking relays need to be slightly delayed w.r.t. reverse-looking relays?
5. In a phase comparison carrier scheme with polarity marks on CTs, prove that
   a. For an internal fault CT secondary currents are in phase.
   b. For an external fault current they are out of phase.
6. What is the effect of carrier failure on
   a. All three carrier-aided distance schemes?
   b. The two unit carrier schemes?

6
Current and Potential Transformers

6.1. INTRODUCTION

Current transformers and potential (voltage) transformers are an integral part of any power system, essentially for measurement and protection. The thermal ratings (i.e., continuous rating without overheating and damaging the insulation) of the current and pressure coils of most measuring devices and protective relays are 110 V line to line for pressure or voltage coils and 0.5 or 1.0 or 5.0 A for current coils or circuits.

Current and potential transformers have two purposes:

1. To reduce line voltage to 110 V and line currents to 0.5 or 1.0 or 5.0 A so that they can safely be handled by measuring devices and protective relays to which they are fed.
2. To physically isolate measuring and protective devices from high power system voltage. This permits easy observation and maintenance of the devices.

In addition to steady-state performance, the performance during and after large input changes is important, in that the output quantity may deviate from a sinusoidal waveform. Many protective relays are required to operate in a time frame shorter than the transient disturbance. Transient performance is therefore highly important.

6.2. POLARITY OR DOT MARKS

Every transformer has *polarity* or *dot* marks. If the currents are assumed to enter the polarity marks of the primary and secondary windings, the ampere-turns of the windings, or the fluxes produced by the windings, are additive. Alternatively, the directions of winding of the primary and the secondary are the same, if viewed from the polarity marks. In other words, since pri-

200 Transformers

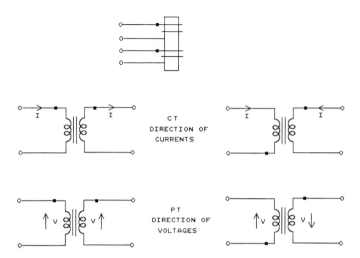

Figure 6.1 Different polarity marks of CT and PT.

mary and secondary ampere-turns for any transformer are always subtractive, if one assumes that current enters the polarity mark of the primary winding then the secondary current must leave the polarity mark.

Figure 6.1 shows the direction of the input and output currents and voltages for CT and PT, with all possible combinations of polarity marks.

6.3. CURRENT TRANSFORMERS

6.3.1. Difference Between Measuring and Protective Devices

The secondary of the measuring CT is connected to the current coils of the ammeter, wattmeter (active power), VAR meter (reactive power), VA (complex power) meter, watt-hour meter (energy), etc. The purpose of these CTs is to measure the system parameters *under normal conditions, not exceeding maximum expected full load*, and not necessarily under fault conditions. The measuring CT is, therefore, required to be accurate to monitor the system quantities and, perhaps, get saturated so as not to reproduce the fault current on the secondary side, to avoid damage to the measuring instruments.

When a fault takes place on a power system, the current tends to increase and voltage tends to collapse. The fault current is abnormal and may be 20 to 30 times the full-load current. Further, it may have a dc offset. For a 5.0-A secondary the short-time fault current could be 100 to 150 A. The CT secondary, however, has a continuous rating of 5.0 A, but it also

Chapter 6

has 100 to 150 A short-time rating. Therefore, the CT secondary will not be damaged.

The protective CT thus has to correctly reproduce this abnormal short-circuit current so that the protective relays operate satisfactorily. The ac component in the fault current is of paramount importance for the relays and has to be reproduced in spite of the dc offset in the primary winding. Thus the protective CT may be not as accurate as the measuring CT, but must work to 20 to 30 times the rated current superimposed on the dc offset.

Figure 6.2 shows two ideal CTs along with a load. Note that the approximate secondary excitation voltage for the two CTs would be

$$V_{excitation} = I_{f,L\,max} R_b\, V \quad \text{(for measuring CT)}$$

$$V_{excitation} = I_{fault} R_b\, V \quad \text{(for protective CT)}$$

where I_{fault} is 20 to 30 times the full-load current.

Thus, it is observed that

$$V_{excitation\ prot.\ CT} = (20\ to\ 30) V_{excitation\ meas.\ CT}$$

Since the magnetic flux is directly proportional to voltage (i.e., $V = 4.44 fN\phi$), the flux in the protective CT would also be 20 to 30 times the flux required for the measuring CT. To summarize: the magnetic flux in the protective CT is very large compared to that in the measuring CT; therefore, the magnetic core size of the protective CT is much bigger than the measuring CT. The measuring CT should saturate at about 1.2 times the full-load current, whereas the protective CT should not saturate up to 20 to 30 times full-load current. Note that the dc offset also needs consideration while designing the protective CT.

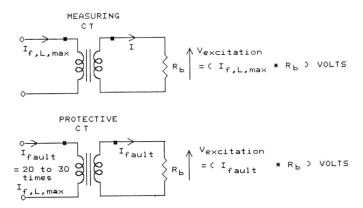

Figure 6.2 Load on measuring and protective current transformer. Protective and measuring CT ratio = 1:1.

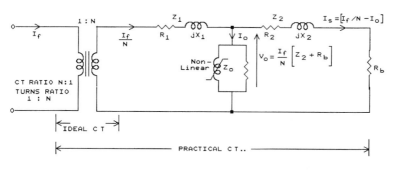

a) IDEAL CT $I_s = I_f / N$
b) PRACTICAL CT $I_s = I_f / N - I_0$
c) V_0 = CT SECONDARY EXCITATION VOLTAGE $\cong \frac{I_f}{N}[Z_2 + R_b]$
   (NEGLECTING I_0)

Figure 6.3 Equivalent circuit of CT as viewed from secondary side.

6.3.2. Equivalent Circuit

The equivalent circuit, as viewed from the CT secondary side for any CT, is shown in Fig. 6.3. In the figure

I_f = line side short-circuit current in amps
$N{:}1$ = current transformer ratio
T = ideal transformer
I_f/N = CT secondary current from ideal transformer
$R_1 + JX_1$ = primary winding resistance and leakage reactance, as viewed from secondary
Z_0 = CT magnetizing impedance, as viewed from CT secondary
$R_2 + JX_2$ = secondary winding resistance and leakage reactance, as viewed from CT secondary
R_B = resistance in ohms of current circuits of relays
$V_0 = (I_f/N)(Z_2 + R_B)$
    = CT secondary excitation or magnetizing voltage
I_0 = CT secondary excitation current at excitation voltage V_0 read from CT secondary V_0 versus I_0 magnetizing characteristics
$I_s = I_f/N - I_0$
    = secondary current delivered to relay burden

Chapter 6

Thus, in an ideal CT, the secondary current is

$$I_s = \frac{I_f}{N}$$

and in a practical CT,

$$I_s = \frac{I_f}{N} - I_0$$

Thus, the practical CT does not reproduce the primary current exactly in amplitude and phase due to its magnetizing current I_0. The magnetizing current is the main source of errors, whether it is a measuring CT or a protective CT.

6.3.3. Steady-State Performance

Figure 6.4 shows the phasor diagram of the current transformer with I_f as the input primary current. Note that $I_s = I_f/N - I_0$. It may further be observed that the magnetizing impedance Z_0 is essentially nonlinear due to nonlinear B-H (due to saturation) characteristics of the magnetic core.

A study of the equivalent circuit will reveal all the properties of the current transformer, namely:

1. Theoretically, the secondary current is independent of the load, except for the magnetizing current.
2. The secondary of the CT must never be open-circuited, since there will be no secondary ampere-turns to balance the primary ampere-turns. The mutual flux will be as shown in Fig. 6.5, resulting in extremely high voltage spikes, thereby damaging the secondary.
3. The ratio and phase angle error depend on the magnetization characteristics, input current and load. The *ratio error* or *ratio correction factor* is the difference between primary and secondary currents, corrected by the nominal CT ratio. *Phase angle error* is the phase difference between primary and secondary currents.

6.3.4. Transient Performance

When a fault occurs the fault current invariably has a dc offset or, in another words, the ac component is superimposed over the dc offset. As we shall see, the effect of dc in the CT primary is to saturate its core, so the CT is not capable of reproducing the ac component on the secondary site until it comes out from saturation.

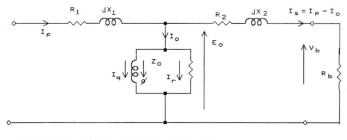

EQUIVALENT CIRCUIT OF 1:1 RATIO CT

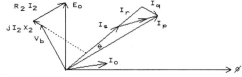

ϕ = MUTUAL FLUX
E_o = SECONDARY INDUCED EMF
V_b = VOLTAGE ACROSS BURDEN $R_b = E_o - I_2 [R_2 + jX_2]$
I_p = PRIMARY CURRENT
I_o = MAGNETISING CURRENT
I_s = SECONDARY CURRENT = $I_p - I_o$
I_r = COMPONENT OF I_o IN PHASE WITH I_s
I_q = QUADRATURE OR MAGNETISING COMPONENT OF I_o
θ = PHASE ANGLE ERROR = ANGLE BETWEEN I_p AND I_s

Figure 6.4 Phasor diagram of current transformer.

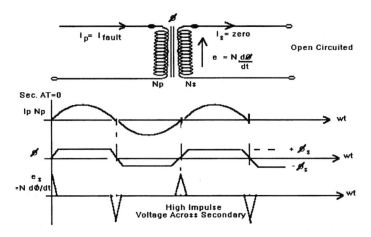

Figure 6.5 CT secondary open-circuited.

Saturation of the magnetic core has tremendous implications on the operation of high-speed protective relays. The high-speed relay needs the ac component in the first one or two power frequency cycles, whereas it may not be available due to saturation. Let us examine the effect of the dc offset on the CT. Refer to Fig. 6.6, wherein it is assumed that the primary current is a decaying dc offset which is being correctly reproduced on the secondary side. The question is, how is the magnetic flux developed?

The fault current, with maximum dc offset, is

$$i = I_m[\underbrace{\sin(wt - 90°)}_{\text{ac component}} + \underbrace{e^{-t/T} \sin 90°}_{\text{dc component}}]$$

Assuming that this primary fault current is being reproduced correctly, the secondary excitation voltage is

$$e_s = I_m R_t[\sin(wt - 90°) + e^{-t/T}]$$

The mutual magnetic flux is given by the equation

$$e_s = N_s \left(\frac{d\phi}{dt}\right)$$

or the flux is

$$\phi = \frac{1}{N_s} \int e_s \, dt$$

where R_t = CT secondary resistance + relay load

$$T = \text{time constant} = \frac{L_L}{R_L} = \frac{X_L}{wR_L}$$

$$= \text{line side time constant}$$

Considering the effect of dc offset alone, we have

$$\text{Flux}_{dc} = \frac{I_m R_t}{N_s} \int_{t=0}^{=t} -Te^{-t/T} \, dt$$

$$= \frac{I_m R_t}{N_s} [-T(e^{-t/T} - 1)]$$

$$\text{Flux} = \frac{R_t I_m}{N_s} T(1 - e^{-t/T})$$

$$\text{Flux}_{dc\,max} = \frac{R_t I_m T}{N_s} \quad \text{as } t \to \infty$$

Thus, the decaying dc offset primary current is to produce an exponentially increasing flux and then remain constant to maximum value. Hence,

At $t = 0$ $\phi = 0$

At $t \to \infty$ $\phi_{max} = \dfrac{I_m R_t T}{N_s}$

Figure 6.6 shows e_s and the flux (integral of e_s) as a function of time. The effect of decaying dc offset primary current (the time constant being $T = X_L/R_L$, i.e., line side parameters on current transformers is to develop the flux from zero to the maximum value of $I_m R_t T \sin(\theta)/N_s$ and then, theoretically, to remain constant.

Now consider only the ac component:

$$e_s = I_m R_t \sin(wt - 90°)$$

$$\text{Flux} = \dfrac{1}{N_s} \int e_s \, dt$$

$$= \dfrac{I_m R_t}{N_s} \int_{t=0}^{=t} e_s \, dt$$

$$= \dfrac{I_m R_t}{w N_s} [-\cos(wt - 90°)]$$

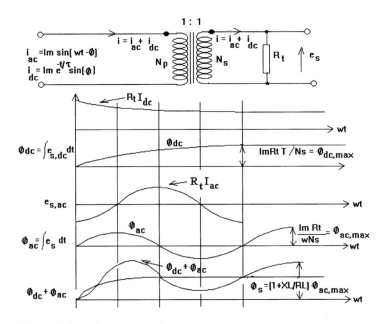

Figure 6.6 Effect of dc offset on CT saturation.

Chapter 6

Therefore,

$$\phi_{ac\,max} = \frac{I_m R_t}{wN_s} = \text{peak value of ac flux}$$

$$\frac{\phi_{dc\,max}}{\phi_{ac\,max}} = \frac{X_L}{R_L} \quad \left(T = \frac{L}{R} = \frac{X_L}{R_L}\right)$$

If it is desired to reproduce the ac component from the primary to secondary side all the time, then Fig. 4.6 shows that the core must be of such a size (fairly large) so as not to saturate: $\phi_s = \phi_{dc\,max} + \phi_{ac\,max}$

$$\phi_{sat} = \left(1 + \frac{X_L}{R_L}\right) \phi_{ac\,max}$$

This equation shows that the magnetic core size increases X_L/R_L times, which is anywhere between 20 to 30 times the design for the peak value of ac flux.

Further, if one expects subsynchronous (<60 or 50 Hz) harmonics in the current, due to a series capacitor in a transmission line and a multimass real model of turboalternator, the flux may further increase. This is due to the fact that flux is inversely proportional to frequency.

6.3.5. Accuracy Class of Current Transformers

The accuracy of any CT is determined essentially by how accurately the CT secondary develops the primary input current. As pointed out earlier, errors can be phase angle or ratio. American and British practices vary in this respect.

6.3.5.1. British Practice for CT Accuracy Class

Class	Ratio error (%)	I_0/I_{sat} (%)	Phase angle error (deg)
S	±3	±3	2
T	±10	±10	6
U	±15	±15	9

where I_0 = magnetizing current and I_{sat} = saturation current. As the primary current, fed to the CT, increases, the secondary excitation voltage increases (it being proportional to secondary current multiplied by load current) and the magnetizing current increases.

The CT secondary excitation characteristics are shown in Fig. 6.7 and are usually plotted as magnetizing current, as viewed from CT secondary, versus secondary excitation voltage.

As the input current increases, the secondary excitation voltage increases and the magnetizing current, more or less, linearly increases, since saturation of the magnetic core has not yet set in. As the current further increases, the CT magnetic core tends to saturate and the magnetizing current increases very rapidly. The knee-point voltage is an indication of CT saturation, beyond which one or both errors, as indicated in the table, increase more than specified.

Another definition of knee-point-voltage is the following: if the secondary excitation voltage increases by 10% over and above the knee-point voltage (V_{knee}), the magnetizing current increases by 50% of the magnetizing current at V_{knee}. V_{knee} is also shown in Fig. 6.7, and it indicates saturation of the CT. The CT should never be required to develop secondary excitation voltage more than V_{knee}. The errors would be beyond these specified of that class.

Thus, the saturation current of any CT is defined as that current, fed to a specified load, beyond which the maximum permissible phase angle and ratio errors are exceeded. The standard saturation factors, per British practice, are 5, 10, 15, 20 and 30.

It is worthwhile to examine the equivalent circuits for an unsaturated CT and a saturated CT. For an unsaturated CT (i.e., secondary excitation voltage < V_{knee}), the magnetizing impedance is fairly large and the exciting current is fairly small. The secondary current is

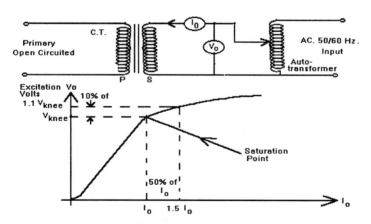

Figure 6.7 Magnetizing characteristics of CT as viewed from secondary.

Chapter 6

$$I_s = \frac{I_p}{N} - I_0$$

For a saturated CT (i.e., secondary excitation voltage $\gg V_{knee}$), the magnetizing impedance Z_0 virtually drops to zero. All the primary current passes through the magnetizing impedance, and the secondary current is zero.

The equivalent circuits of the unsaturated and saturated CTs are shown in Fig. 6.8. Note that for the saturated CT the magnetizing impedance is shunted by a short circuit. The input impedance, as viewed from the secondary, is only Z_s.

What happens if the CT secondary load is not purely resistive (i.e., unity power factor)? Instead of a mathematical proof, we use physical interpretation, which is more appealing. Figure 6.9 shows a faulted line and the CT load has the same power factor as that of the line fault current. It is clear from symmetry on the line and CT secondary sides that the CT secondary excitation voltage will be purely sinusoidal, similar to the source voltage on the line side, even if the line current has a dc offset. This load is called the *line impedance replica*. In effect, if the CT secondary load is

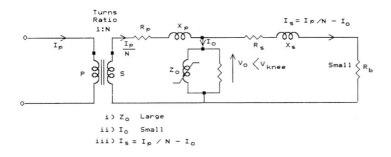

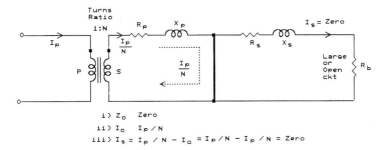

Figure 6.8 Equivalent circuits for saturated and unsaturated current transformers.

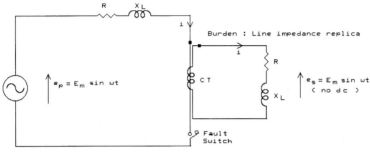

i) Line side voltage $e_p = E_m \sin \omega t$
ii) Line impedance = Burden impedance
iii) $e_s = E_m \sin \omega t$
iv) No dc flux in the CT

Figure 6.9 Replica impedance as load on CT.

matched to the system side load, the CT secondary excitation voltage is purely sinusoidal and the saturation is nil or a minimum.

6.3.5.2. American Practice

In the United States the ratio error is called the *ratio correction factor* (RCF) and is defined as

$$\% \text{ error} = \frac{NI_s - I_p}{I_p} \times 100$$

where N = secondary turns/primary turns
 = nominal CT ratio
I_s = secondary current
I_p = primary current

The code 2.5 L/H 250 means the percentage accuracy is 2.5% at $V_{\text{knee}} = 250$ V, L is the low-leakage toroidal CT and H is the high-leakage wound-type CT. In practice, the ratio error correction is provided by a few additional terms on the secondary side.

6.4. POTENTIAL TRANSFORMERS

6.4.1. Introduction

The purpose of a potential (voltage) transformer is to physically isolate the relay equipment from high voltage and to reduce the voltage to 110 V line to line. Standard international practice is to design the pressure or voltage circuits to 110 V line to line or $110/\sqrt{3}$ phase-to-neutral voltage for the relays.

6.4.2. Steady-State Performance

Figure 6.10 shows the phasor diagram for the PT. The percent ratio error in these voltage transformers is defined as

$$\% \text{ ratio error} = \frac{NV_{sec} - V_{pri}}{V_{pri}} \times 100$$

where N = primary turns/secondary turns
= nominal voltage ratio
V_{pri} = applied primary high voltage
V_{sec} = secondary voltage

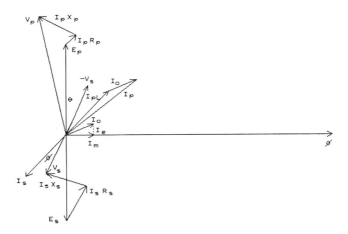

i) $\emptyset$: Mutual Flux
ii) E_p : Primary Induced Voltage, E_s : Secondary Induced Voltage
iii) V_p : Primary Applied Voltage, $\emptyset$: Secondary P.F.
iv) I_p : Primary Current, I_{pL} : Load Component Of I_p
v) $I_p R_p$: Primary Resistive Drop, $I_s R_s$: Secondary Resistance Drop
vi) $I_p X_p$: Primary Reactance Drop, $I_s X_s$: Secondary Reactance Drop
vii) I_o : Magnetising Current
viii) I_m, I_e : Magnetising And Loss Component
ix) θ : Phase Angle Error

Figure 6.10 Phasor diagram of potential voltage transformer.

The limits of voltage and phase angle errors according to British practice are:

British Standards BS 3941:1965

Class	0.9 to 1.1 times rated voltage 0.25 to 1.0 times rated load at unity power factor	
	Ratio error (%)	Phase angle error (min)
A	±0.5	±20
B	±1.0	±30
C	±2.0	±60

For protective purposes, the accuracy is important down to small voltages, about 5%, due to its collapse on fault.

	0.25 to 1.0 times rated load at unity power factor			
	0.05 to 0.9 times rated primary voltage		1.1 to V_{max} times primary voltage	
Class ↓	% voltage error	Phase error (min)	% voltage error	Phase error (min)
E	±3	±120	±3	±120
F	±5	±250	±10	±300

The types of PTs are *magnetic core*, used to 132 kV, and CVT (capacitor-type voltage transformers), used above 132 kV. Magnetic core PTs have increased error at voltages down to 1.0%, which is very likely, due to collapse of voltage on faults. This inaccuracy is essentially due to the non-linear B-H curve at low magnetizing ampere-turns. The errors are tolerable and normally ignored.

The standard construction technique of the CVT is shown in Fig. 6.11. In the circuit capacitors C_1 and C_2 are used as potential dividers, and the voltage across C_2 is brought down to about 10% of the line voltage. This voltage is further fed via inductor L to the conventional magnetic type of PT. The PT secondary delivers 63.5 V, phase to neutral, for use in relays.

Chapter 6

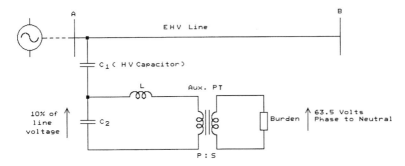

Figure 6.11 CVT (capacitor-type voltage transformer).

The inductance L is inserted to reduce errors when the PT load and the frequency are varied. The inductor compensates for the capacitive source impedance (parallel combination of C_1 and C_2.

Recall that we do have subsynchronous (below 60 or 50 Hz) frequencies in series-compensated lines, with a multimass model of turboalternators.

Invariably, one uses an electronic amplifier, feeding to static or processor-based relays, to reduce the load on the CVT and, thus, reduce the phase angle and ratio errors.

6.4.3. Transient Performance

The transient response of a conventional PT is much superior to the CVT. Inductance is the only energy storage element, whereas in CVT one has capacitance as well as inductance. This results in an oscillatory solution for CVT, in the event of any sudden changes on the line side.

The exciting impedance Z_0 of the auxiliary magnetic PT and the capacitance of the potential divider together form a resonant circuit, which will normally oscillate at subsynchronous frequency (<60 or 50 Hz). It has

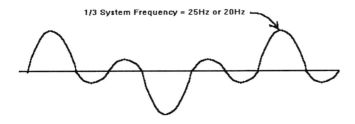

Figure 6.12 CVT subsynchronous oscillations.

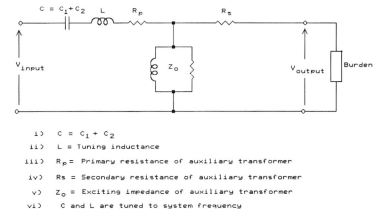

Figure 6.13 Equivalent circuit of CVT.

been reported that, due to resonance, the rms value of subsynchronous oscillations may rise to 25% to 50% of normal voltage. The typical waveform of these oscillations is shown in Fig. 6.12

Transient performance can be calculated from the equivalent circuit shown in Fig. 6.13. At the moment, no particular standards exist that define the transient behavior of the voltage transformers; if a relay malfunctions, each case needs to be investigated.

6.5. PROBLEMS AND EXERCISES

1. What is the basic difference between measuring and protective CTs, as far as saturation flux density is concerned?
2. Prove that the peak value of flux in a CT, for input fault current with decaying dc offset, is

$$\text{Peak value of flux} = \left(1 + \frac{X_L}{R_L}\right) \times \text{peak value of ac flux}$$

where X_L = line reactance and R_L = line resistance.
3. Intuitively, one says that the transient behavior of a CVT is no better than a conventional magnetic core PT. State why.
4. A 100:5 A current transformer is fed on the primary side by 2000 A. The secondary lead resistance is 1.0 ohm, whereas the relay coil resistance is 2.0 ohms. The secondary leakage reactance may be ignored. The magnetizing characteristics may be assumed linear

up to V_{knee} V, and the magnetizing current is 5.0 mA/V. After the excitation voltage goes beyond V_{knee}, the magnetizing current may be assumed to be 20.0 mA/V. Calculate the CT secondary excitation voltage and the ratio error for

a. $V_{knee} = 300$ V
b. $V_{knee} = 250$ V

Draw conclusions, if any.

7
Basics of Differential Relays

7.1. INTRODUCTION

As the name implies, the differential *relay* checks the difference between the input and output currents for any power system element, either in amplitude or in phase or both, to determine whether the state of the power system element is health or faulty. In the event of a substantial difference, the element is assumed to be faulty and the circuit breakers are tripped to disconnect the element from the remaining healthy system.

In most conventional differential schemes the input/output currents are checked for any difference in their phasor form (amplitude as well as phase), and a decision is made regarding the faulty or healthy state of the power system element.

As an exception, the phase comparison carrier scheme for EHV lines (covered in Chapter 5) checks the phase difference between the input and output currents in phase only to decide the state of the EHV line. It is a kind of differential scheme, but the protection engineers do not like it to be so called.

All differential schemes generate a protective zone, which is accurately delimited by the input/output current transformer locations. Thus, these schemes are unit schemes and are incapable of providing backup protection.

7.2. SIMPLE DIFFERENTIAL SCHEME

To explain the basic principle of a simple differential relay, we consider a single-phase power transformer of ratio 1:1 energized from one end only. Refer to Fig. 7.1 and note carefully the polarity marks on the power and current transformers. Since the power transformer voltage ratio is assumed to be 1:1, the required CT ratios at both ends need to be 1:1 also.

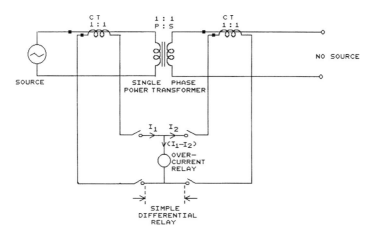

Figure 7.1 Simple differential scheme.

7.2.1. Behavior on Load Current

Figure 7.2 shows the combination of the two CTs and an instantaneous over-current relay, connected in the difference or spill circuit, as the constituents of the simple differential scheme. The normally open contacts of this OC relay are wired to trip the circuit breakers.

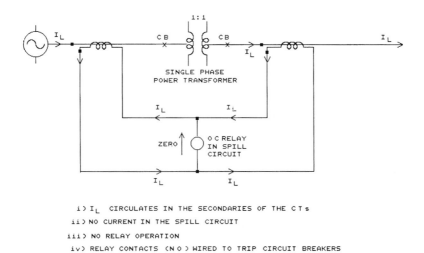

Figure 7.2 Behavior on load current.

Chapter 7

Note that the input and output currents of the power transformer will be more or less the same in magnitude and phase. The slight difference is due to additional magnetizing current on the primary side. It is ignored for the moment.

Let I_L be the secondary load current. The primary current is also approximately I_L. The secondaries of the input/output CTs faithfully (neglecting CT errors) reproduce these load currents. Note carefully the polarity marks on the two CTs.

The secondary currents, under normal power flow, will circulate among the two CTs, and there will be no current in the spill or difference circuit, where the instantaneous OC relay is connected. The relay does not operate, so the circuit breakers don't trip. Thus, the transformer is not faulty. Since the currents circulate in the CT secondaries, this scheme is called the *Merz-Price circulating differential scheme*.

7.2.2. Behavior on External Fault

Figure 7.3 shows an external or through fault. A through fault occurs outside the CT locations and not in the power transformer. Denote this current by $I_{f,\text{ext}}$; it is the same on both sides of the transformer, since the transformer ratio is 1:1.

As with normal power flow, the current in the OC relay is zero. The external fault current circulates among the two CT secondaries. The OC relay does not operate. Rightly, we do not want to isolate the transformer.

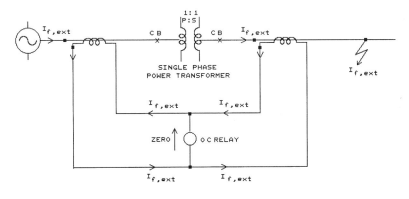

Figure 7.3 Behavior on external fault.

The fault is external and not within the transformer. Thus, the operation of this simple differential relay is correct.

7.2.3. Behavior on Internal Fault

Figure 7.4 shows a fault in the transformer. The fault current is contributed from the source side, which is end A described in the figure. No fault current is contributed from right end B, since there is no source of power.

Denote the fault current by $I_{f,int}$. CT_A secondary delivers the fault current, but CT_B does not deliver any current. The current distribution shows that the OC relay in the spill or difference circuit if the fault current $I_{f,int}$ itself. Thus, the relay operates and deenergizes the transformer from the source. This is the correct operation of the differential relay. To summarize: The differential relay does not operate on normal power flow or an external or through fault in an adjoining power system element. It rightly operates on an internal fault, thereby deenergizing the faulty transformer from the power sources.

7.2.4. Relay Characteristics on $(I_1 + I_2)/2$ Versus $I_1 - I_2$ Diagram

Figure 7.5 shows that, whatever the current $(I_1 - I_2)/2$ is, the relay operates whenever the current in the OC relay exceeds I. The characteristic is therefore a horizontal line displaced vertically by I_{pu}.

7.2.5. Stability Ratio and Limitations

7.2.5.1. Introduction

The differential relay should

1. Not operate on power flow and external fault
2. Operate on an internal fault.

Both external and internal fault currents depend on the fault type (three phase, L-L-G, L-L, L-G), fault location and possible variation in source impedance (changes in the strength of the power source, such as increase or decrease in generation, lines in or out behind the transformer under protection, etc.).

Recall that

1. The fault current is minimal under the following conditions (Fig. 7.6):
    a. The source impedance is maximal.

Chapter 7

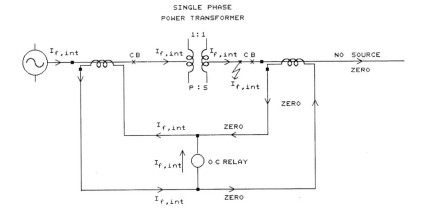

i) FOR INTERNAL FAULT CURRENT IN O C RELAY = $I_{f,int}$..
ii) RELAY AND C B s TRIPPED (CORRECT OPERATION)

Figure 7.4 Behavior on internal fault.

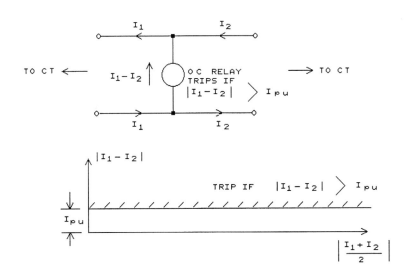

i) SIMPLE DIFFERENTIAL RELAY OPERATES FOR DIFFERENCE BETWEEN I_1 AND I_2
ii) THE OPERATION IS IRRESPECTIVE OF $|I_1+I_2/2|$

Figure 7.5 Differential relay characteristics.

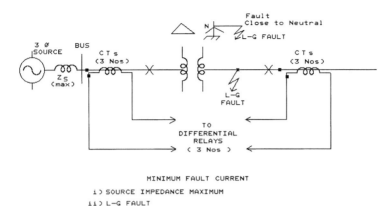

Figure 7.6 Conditions for minimum fault current.

    b. The fault is L-G.
    c. The L-G fault is close to the grounded neutral.
2. The fault current is maximal under the following conditions (Fig. 7.7):
    a. The source impedance is minimal.
    b. The fault is three phase.
    c. The three-phase fault is close to the transformer terminals.

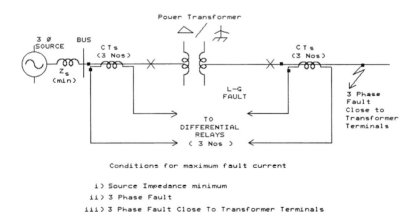

Figure 7.7 Conditions for maximum fault current.

Chapter 7

The *stability ratio*, S, of any differential scheme is defined as

$$S = \frac{I_{f,\text{ext,max}} \text{ (no malfunction)}}{I_{f,\text{int,min}} \text{ (relay operates)}}$$

where $I_{f,\text{ext,max}}$ = maximum external fault current within which the relay will not malfunction; if exceeded, the relay will malfunction
$I_{f,\text{int,min}}$ = minimum internal fault current for which the relay will correctly operate; if fault current is less the relay will not operate, which is not desirable.

To find the stability ratio, one requires relay trip, and internal and external fault characteristics.

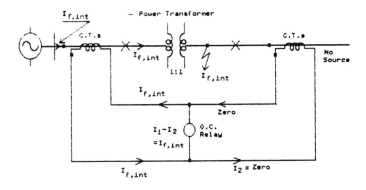

i) $I_1 = I_{f,\text{int}}$
ii) $I_2 = \text{Zero}$
iii) $(I_1+I_2/2) = I_{f,\text{int}}/2$
iv) $I_1-I_2 = I_{f,\text{int}}$
v) slope $= I_1-I_2/[I_1+I_2/2] = \dfrac{I_{f,\text{int}}/2}{I_{f,\text{int}}} = 2$

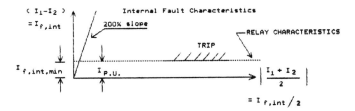

Figure 7.8 Internal fault characteristics.

7.2.5.2. Internal Fault Characteristics

Figure 7.8 shows an internal fault. Note that

$$I_1 = I_{f,int}$$
$$I_2 = 0$$
$$\frac{I_1 + I_2}{2} = \frac{I_{f,int}}{2}$$
$$I_1 - I_2 = I_{f,int}$$
$$\text{Slope} = 2.0$$

Thus, the internal fault characteristics have a 200% slope (i.e., slope of 2.0), as shown in Fig. 7.8. Now superimpose the relay characteristics and the point of intersection will give the minimum fault current required for the relay to operate ($I_{f,int,min}$).

7.2.5.3. External Fault Characteristics

Figure 7.9 shows an external or through fault. Recall that the current in the spill circuit is zero and the relay, rightly, does not operate. However, there is current in the spill circuit, which increases as the through-fault current increases, causing the relay to malfunction. This, is essentially due to the unequal magnetizing currents of the two CTs. Such current is usually called CT *unbalance current* for through faults.

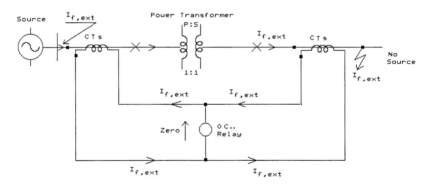

 i) No Current in OC Relay
 ii) No Operation Of Relay
 iii) No Tripping Of CBs On External Fault
 iv) Correct Operation

Figure 7.9 External fault.

Figure 7.10 shows the differential relay around the power transformer with the equivalent circuits of the CTs and the OC relay in the spill circuit. Note that

$I_{f,ext}$ = external fault current
Z_{1A} = CT A primary impedance
Z_{0A} = CT A magnetizing impedance
Z_{2A} = CT A secondary impedance
I_{0A} = CT A magnetizing current
V_{0A} = CT A secondary excitation voltage
Z_{1B} = CT B primary impedance
Z_{0B} = CT B magnetizing impedance
Z_{2B} = CT B secondary impedance
I_{0B} = CT B magnetizing current
V_{0B} = CT B secondary excitation voltage
R_{lead} = CT lead resistances

Thus, the current in the spill circuit or in the instantaneous OC relay is no longer zero but the difference in the magnetizing currents of the two CTs. The differential scheme malfunctions if

$$I_1 - I_2 = I_{0B} - I_{0A} > I_{pu}$$

Such a malfunction on an external fault is undesirable and needs further investigation.

The magnetizing currents of the CTs are the functions of CT secondary excitation voltages and can be read from the CT secondary magnetization characteristics. The CT secondary excitation voltages, neglecting the magnetizing currents, are (Fig. 7.5)

$$V_{0A} = I_{f,ext}(Z_{2A} + R_{lead})$$
$$V_{0B} = I_{f,ext}(Z_{2B} + R_{lead})$$

For simplicity let the CTs be similar and $V_{0A} = V_{0B}$. The nonlinear magnetization characteristics of the CTs are shown in Fig. 7.11, and the hatched region shows how the difference in magnetizing currents $(I_{0A} - I_{0B})$ increases as the excitation voltage increases. This difference is basically due to different CT saturation levels. Note that the excitation voltage is proportional to $I_{f,ext}$.

Differential Relays

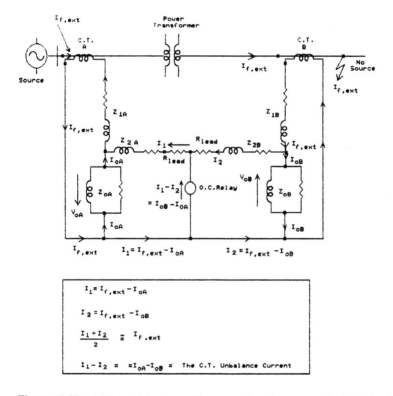

Figure 7.10 Differential relay with current transformer equivalent circuit.

Referring back to Fig. 7.11 we have

$$I_1 = I_{f,\text{ext}} - I_{0A}$$
$$I_2 = I_{f,\text{ext}} - I_{0B}$$

Therefore,

$$\frac{I_1 + I_2}{2} = I_{f,\text{ext}} - \frac{I_{0B} - I_{0A}}{2}$$

$$\approx I_{f,\text{ext}}$$

$$I_1 - I_2 = I_{0A} - I_{0B}$$

$$= \text{CT unbalance current}$$

Figure 7.12 shows the external fault and relay trip characteristics. The point of intersection gives the maximal external fault current, $I_{f,\text{ext,max}}$, beyond which the relay malfunctions.

Chapter 7

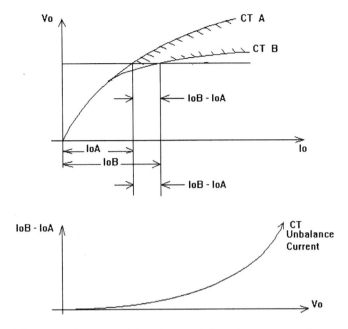

Figure 7.11 Magnetizing characteristics of current transformers.

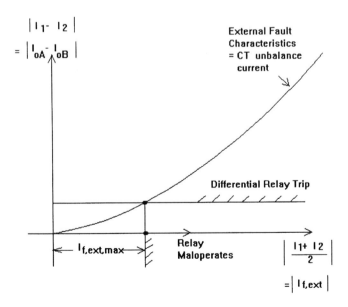

Figure 7.12 External fault characteristics.

Differential Relays

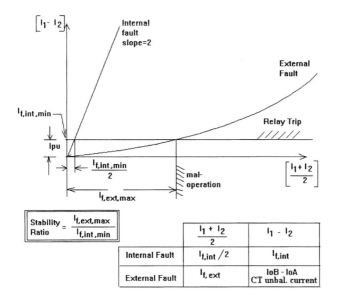

Figure 7.13 Relay/internal fault/external fault characteristics and stability ratio.

7.2.5.4. Stability Ratio and Limitations

To summarize, the relay trip characteristics, the internal fault characteristics, the external fault characteristics, the minimum internal fault current for which the relay operates satisfactorily and the maximum through-fault current beyond which the relay malfunctions are given in Fig. 7.13. The stability ratio, $I_{f,ext,max}/I_{f,int,min}$, should be as high as the design permits. The desired values vary from 100 to 300. For the simple differential relay the stability ratio is extremely poor, so a biased differential relay was developed.

7.3. BIASED OR PERCENTAGE DIFFERENTIAL RELAY

7.3.1. Introduction

The simple differential relay is unable to accommodate large external fault currents without malfunctioning. Therefore, to improve relay operation, it was restrained by inserting restarting coils and energizing them with through-fault current. This additional restraint is called *biasing* the relay.

Fig. 7.14 shows a biased differential relay with two restraining coils and one operating coil on the electromechanical balanced beam structure. The relay operates if the operating force is more than the restraining force:

Chapter 7

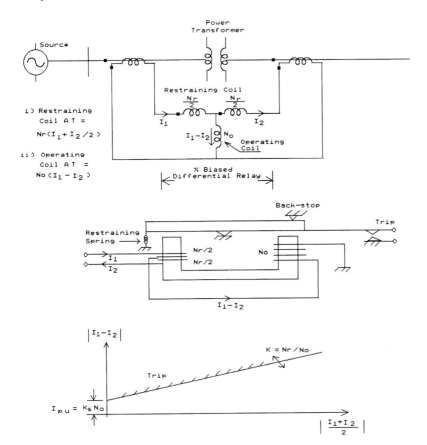

Figure 7.14 Biased differential relay characteristics.

$$|(I_1 - I_2)N_0|^2 > \left|\frac{(I_1 + I_2)N_r}{2}\right|^2 + K_s$$

or

$$|I_1 - I_2| > K\left|\frac{I_1 + I_2}{2}\right| + \frac{K_s}{N_0}$$

where

$$\% \text{ Bias} = \% \text{ Slope} = K = \frac{N_r}{N_0}$$

This characteristic is shown in Fig. 7.14.

Differential Relays

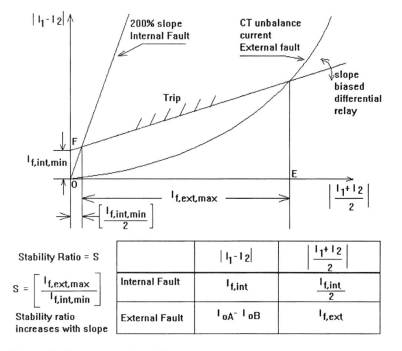

Figure 7.15 Increased stability ratio for biased differential relay.

7.3.2. Stability Ratio

The increase in maximum through-fault current that the relay can accommodate without malfunctioning is shown in Fig. 7.15. It is a function of the slope. Note that the minimum internal fault current for which the relay operates satisfactorily does not deteriorate much.

$$\text{Stability ratio} = \frac{I_{f,\text{ext,max}}}{I_{f,\text{int,min}}} = \frac{OE}{OF} \to \text{large}$$

Thus, the biased differential relay is very stable on through faults and is universally used to protect alternators and transformers.

7.4. VARIABLE-SLOPE OR PIECEWISE LINEAR DIFFERENTIAL RELAY

Theoretically speaking, the ideal biased differential relay should have tripping characteristics slightly above the CT unbalance current (i.e., external

Chapter 7

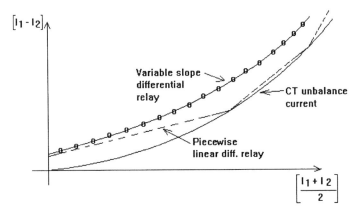

i) No intersection of relay characteristics with CT unbalance current

ii) $I_{f,ext,max}$ = Infinity

iii) Stability ratio = S = Infinity

Figure 7.16 Variable slope of piecewise linear differential relay.

fault characteristics). Thus, the ideal biased differential relay is the variable-slope type or piecewise linear, as shown in Fig. 7.16. Such a relay has an infinite stability ratio.

Not many attempts have been made to generate such nonlinear (i.e., variable slope) relay characteristics, presumably due to variations in CT unbalance currents at different locations. Perhaps a processor-based relay could adjust the slope on site and resolve the problem. The matter requires further investigation and acceptability by the electric utilities.

7.5. PROBLEMS AND EXERCISES

1. Prove that the CT unbalance current for a through fault is equal to the difference between the magnetizing currents of the input and output CTs.
2. In a differential scheme the input/output CTs have the same secondary excitation voltage. The CTs have dissimilar open-circuit (V_0 versus I_0) characteristics. Draw these characteristics and prove that as the CT secondary excitation voltage increases, the CT unbalance current increases rapidly.

3. For an internal fault with a source at one end only, prove that the slope is 200% on an $(I_1 + I_2)/2$ versus $I_1 - I_2$ diagram.
4. Draw the 25% biased differential relay characteristics, internal fault characteristics and the approximate CT unbalance (for through fault) characteristics. Find the stability ratio.
5. Justify the necessity of a variable slope differential relay with on-sight changes in the slope.

8
Generator Protection

8.1. INTRODUCTION

A modern turboalternator with two control systems is shown in Fig. 8.1. The figure shows a steam turbine driving a three-phase alternator. The automatic controls are (a) a speed-governing mechanism to keep the frequency reasonably constant (frequency is directly proportional to speed), and (b) an automatic voltage regulator to keep the terminal voltage reasonably constant. Such a turboalternator could be subjected to

1. Faults in the three-phase stator winding
2. Faults in the rotor (i.e., field winding)
3. Faults in the prime mover

We shall classify the faults, the possible damage if no corrective action is taken and the protective devices that are extensively applied.

8.2. TYPES OF FAULTS AND THEIR DETECTION
8.2.1. Stator Faults

The stator core has a three-phase star winding with the neutral resistance-grounded through the distribution transformer. The grounding resistance limits the ground-fault current to approximately the full-load alternator current. This avoids extensive damage to the stator laminations at the fault point. Such damage requires dismantling the generator and rebuilding the magnetic core, requiring equipment to be shut down for several days.

The three-phase stator winding is subject to the following faults (the stator core is solidly grounded):

1. Three-phase, L-L-G, L-L and L-G faults in the slots of stator core, up to 10 possible faults

234   **Generator Protection**

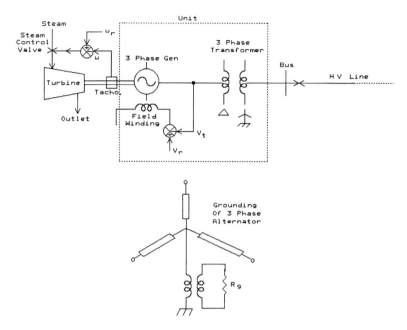

Figure 8.1   Modern turboalternator.

2. Interturn faults on the same phase winding
3. Unbalanced stator currents, either due to severe load unbalance or closing or opening of only two contacts of the circuit breaker

8.2.1.1.   Faults in the Stator Winding

Figure 8.2 shows three biased differential relays and six current transformers. In each phase winding the current is the same at the neutral and terminal ends, so the CT ratios must be identical. One or more relays operate to detect the various types of faults.

All three relays need to be wired to deenergize the field winding and the outgoing ac circuit breaker. Deenergizing the field cuts off the short-circuit current by suppressing the internal generated voltage. Tripping the outgoing CB cuts off the current fed from the grid. Finally, the turbine governor, etc., ensures complete shutdown of the boiler.

Since all six CTs have identical ratios, they are identical in all other respects. Thus, for through or external faults the CT unbalance current is relatively small. The slope of the percentage differential scheme is therefore small, around 10% to 25%. The pickup values are about 0.15 to 0.5 A (for

Chapter 8 235

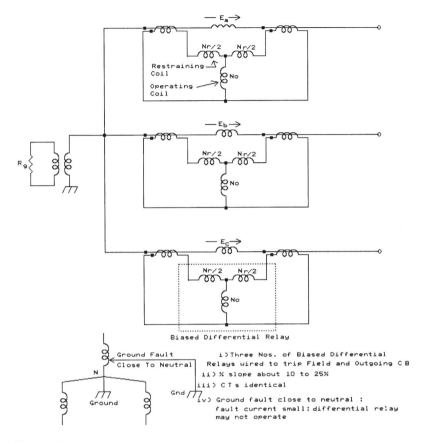

Figure 8.2 Biased differential relays for generator faults.

a 1.0-A relay). The operating times are 80 to 180 msec for the electromechanical relay and 25 to 35 msec for the static relay.

It may further be noticed from Fig. 8.3 that if the ground fault is very close to the neutral, the fault current is too small to be detected by a differential scheme. Therefore, an additional ground-fault relay needs to be connected, as shown.

The figure shows a generator feeding a delta/star step-up transformer. This is the normal practice and is called a *unit type* connection. An overvoltage relay is connected to detect the ground fault, which essentially responds to a shift of the generator neutral with respect to ground. A third harmonic filter is employed so that the relay responds to only the fundamental frequency. Every phase of the generator generates a small amount

236   Generator Protection

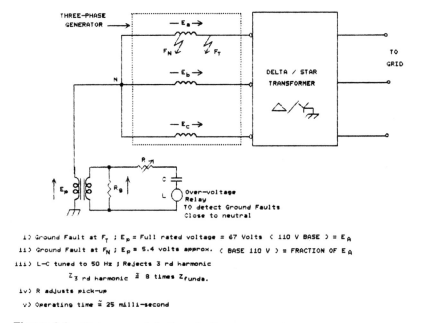

i) Ground Fault at F_T ; E_p = Full rated voltage = 67 Volts (110 V BASE) = E_A
ii) Ground Fault at F_N ; E_p ≈ 5.4 volts approx. (BASE 110 V) = FRACTION OF E_A
iii) L-C tuned to 50 Hz ; Rejects 3 rd harmonic
              $Z_{3\ rd\ harmonic} \cong 8\ times\ Z_{funda}$.
iv) R adjusts pick-up
v) Operating time ≅ 25 milli-second

Figure 8.3   Stator ground-fault protection.

of third harmonics, which combine and pass through the grounding resistance. Hence, a third harmonic filter is needed.

The ground-fault relay can withstand a continuous 67.0 V (in the event of a ground fault close to the generator terminal) and its pickup value is around 5.4 V (8% of 67 V). The operating time can vary from 25 msec to about 4.0 sec. The tuned filter formed by the relay coil and the external capacitor rejects third harmonic currents.

8.2.1.2. Interturn Fault on the Same Phase

Figure 8.4 shows that for an interturn fault the input and output currents on the faulted phase are the same. The differential relay does not operate. To detect an interturn fault on the same phase, one has to use split-phase winding in the alternator. Each phase has two windings connected in parallel (Fig. 8.5). These windings will not share the current equally (i.e., I_L) if the turns on one of the windings becomes shorted. The protection applied is either a transverse differential scheme or a core-balance scheme. The basic principle is that the difference in the two split windings drives an instantaneous OC relay. Both are shown in the figure. The corrective action taken

Chapter 8

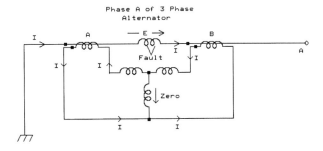

i) Inter-turn fault on the same phase
ii) End A and end B current same
iii) Conventional differential scheme fails to operate

Figure 8.4 Failure of differential relay on interturn fault of the same phase.

by these relays is the same as tripping the field breaker and the outgoing circuit breaker.

8.2.1.3. Unbalanced Stator Currents

The stator currents could be a set of unbalanced currents, due either to load unbalance or because the CB is not closing or opening one of the poles. The stator currents now have plenty of negative sequence component, thus producing a field rotating in the opposite direction to that of the field structure. The field and damper windings will have double the normal power frequency components (i.e., 100 or 120 Hz) of eddy currents, thereby causing the rotor to overheat. This does not immediately trip the alternator, since it has enough thermal capacity to withstand this current for some time. The negative sequence current withstand capability, in seconds, is

$$T = \frac{K}{I_2^2}$$

where

$K_{\text{relay side}} = T = 5.0$ for large turboalternators

$\phantom{K_{\text{relay side}} = T} = 90$ for low MVA alternators

As an example, for a 1000-MVA, 13.2-kV, three-phase turboalternator let $K = 10$ sec. Then:

1. $I_{\text{full load}} = \dfrac{1000 \text{ MVA}}{\sqrt{3} \times 13.2 \text{ kV}} = 43{,}734 \text{ A}$

238 **Generator Protection**

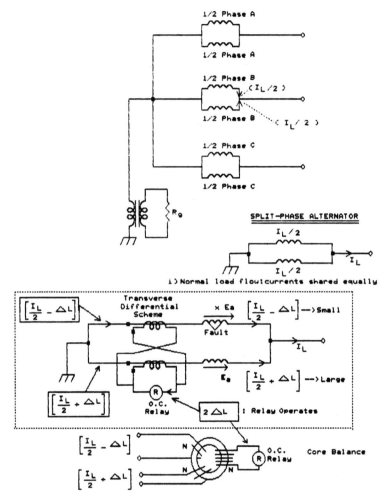

Figure 8.5 Split-phase winding alternator and transverse or core-balance differential scheme.

2. Let CT ratio = 50,000/5.0 A.
3. The static version of the negative sequence relay has continuously adjustable I_{pu} of 0.1 to 1.0 A.
4. Let the anticipated minimum value of the negative sequence current be 15% of the full-load current. Therefore,

$$I_{2,\text{line side}} = I_{\text{full load}} \times 0.15 = 50{,}000 \times .15$$
$$= 7{,}500 \text{ A}$$
$$I_{2,\text{relay side}} = \frac{7500/5.0}{50{,}000} = 0.75 \text{ A}$$

5. Thus, choose $I_{pu} = 0.75$ A.
6. Solve for time:

$$T = \frac{K}{I_2^2}$$
$$= \frac{10.0}{(0.75)^2}$$
$$= 18.0 \text{ sec} \quad \text{for 3.25 A of positive current}$$

7. Adjust the time dial so that the operating time is $T = 18.0$ sec.

Figure 8.6 shows the ability of alternators to withstand negative sequence current along with the available relay characteristics.

8.2.2. Rotor Faults

The rotor of the modern turboalternator is invariably a round-rotor type having a field winding. The field winding can experience (a) a short-circuit between the field winding and the rotor core, resulting in a ground fault, or (b) a loss of field excitation, causing the machine to run as an induction generator.

8.2.2.1. *Ground Faults*

If the field winding is not grounded, a ground fault is virtually no fault. The field excitation remains the same as prior to fault, and the operation of the turboalternators is unaffected. Unfortunately, another subsequent ground fault on the field winding will shunt a certain number of field winding turns. this causes asymmetry of the field magnetomotive force (MMF) on the rotor, causing excessive rotor vibration and much damage. A ground-fault relay, described in the following, is required to trip out the generator. This is shown in Fig. 8.7.

It is desirable to ground the field winding at a suitable point so that the first ground fault can be detected and corrective action taken before a second damaging fault takes place. The protection is shown in Fig. 8.7. A nonlinear resistor in series with resistors R_1 and R_2 are paralleled with the

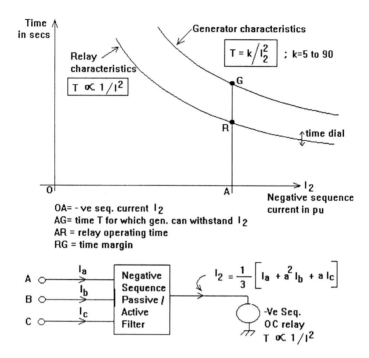

Figure 8.6 Negative sequence current withstand capability and the negative sequence relay.

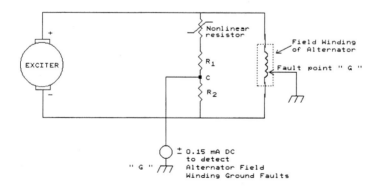

i) Point C, called the null point, is at mid-point potential of field winding when exciter voltage is 100%

ii) Nonlinear resistor varies this null-point to ensure rotor ground fault detection, anywhere on the field winding

Figure 8.7 Ground-fault protection of rotor.

Chapter 8

field winding. The connection between R_1 and R_2 is grounded through a bipolar moving-coil-type relay set to around ± 0.15 mA.

8.2.2.2. Field Failure or Loss of Excitation

In Fig. 8.8 the turboalternator is assumed to be delivering the following complex power S_g in MVA:

$$S_g = P_g + JQ_g$$

Assuming complete failure of the excitation (i.e., $I_f = 0$ as $t \to \infty$) and neglecting all losses, we have

$$S_g = P_g - JQ_g$$

Note the reversal of reactive power. Clearly, the alternator has a prime-mover mechanical intput, whereas the field excitation is lost. This results in the following

1. The alternator runs as an induction generator with greater than synchronous speed.

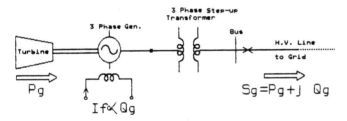

i) Generation intact
ii) Delivering reactive power Q_g

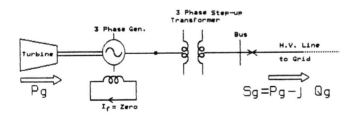

i) Loss of excitation
ii) Absorbs reactive power $-Q_g$

Figure 8.8 Field failure or loss of excitation.

2. The alternator was delivering reactive power to the grid to the extent of Q_g. This reactive power is now absorbed by the alternator to produce the mutual flux. This results in the alternator delivering $-JQ_g$.

The values of R and X, as functions of active and reactive power, are

$$R = \frac{V^2 P}{P^2 + Q^2}$$

$$X = \frac{V^2 Q}{P^2 + Q^2}$$

Thus, the postfault value of X changes from positive to negative, whereas the postfault value of R is unaltered. This is not strictly true, since the friction, windage and copper losses in the stator winding tend to increase.

The conversion of the alternator to an induction generator takes several seconds. Similarly, the losses in friction, windage and stator winding increase, resulting in the locus of impedance viewed from the machine terminals coming closer to the R-axis (i.e., increased P results in decreased R). This locus, as a function of time, is shown in Fig. 8.9, by locus AG.

The loss-of-excitation (LOE) relay should not malfunction on a stable power swing after clearing a three-phase fault on the HV side of the step-

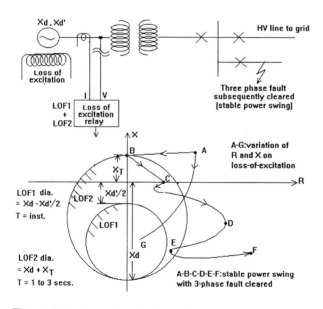

Figure 8.9 Loss of excitation relay.

up transformer. The fault is immediately after the unit transformer, and then cleared. This is also shown in Fig. 8.9, along with the locus A-B-C-D-E-F. Therefore, the LOE relay should operate on loss of excitation only and must discriminate (not malfunction) on the stable power swing from which the machine is recovering. The LOE relay, shown in Fig. 8.9, has dual characteristics. The first is a circle having diameter $X_D - X'_D/2$ a distance $X'_D/2$ from the origin of the R-X diagram. It operates more or less instantaneously. The second characteristic is also a circle, with diameter $X_D + X_T$, but it operates with an intentional delay, adjustable from 1.0 to 3.0 sec, to tide over the possible malfunction on a stable power swing. This type of dual characteristic is widely applied.

8.2.3. Overload and Overheating Protection

All alternators are subjected to overload and, hence, overheating. Therefore, an overload relay is required to give an alarm for suitable corrective action, such as load dropping, etc., to be taken. The relay is not necessarily wired to trip the circuit breaker. The alternators are equipped with embedded resistance sensors in a bridge circuit (i.e., wheatstone). When the temperature exceeds 120° C, the bridge becomes unbalanced, and the relay across this unbalance gives an alarm.

8.2.4. Overspeed Protection

When a load on the alternator is dropped, the speed increases. The whole dynamics of speed depends on prime-mover input, the machine inertia constant and the action of the speed-governing mechanism. In any case, we need an overspeed device for speed control and an alarm which could be either of the following:

1. Overfrequency relay, the frequency being directly proportional to speed.
2. A permanent magnet tachogenerator mounted on the shaft. The generated voltage is now directly proportional to speed.

8.2.5. Loss of Prime-Mover Input (Motoring of Turboalternator)

A turboalternator whose excitation is intact runs as a synchronous motor when there is a failure of prime-mover input (prime-mover input is the active power, minus the losses, delivered to the grid). The alternator, instead of delivering any active power, runs as a synchronous motor and draws very small active power to overcome the friction and windage losses of the combined generator and prime mover.

The prime mover does have small steam/water drops which now churn, damaging the turbine blades. Hence a relay is required. The relay is a simple reverse power relay adjusted to about 3% to 5% of the rated nameplate power (in megawatts) of the machine.

8.2.6. Out-of-Step Conditions

A synchronous machine is subjected to two types of power swings:

1. Stable power swing from which the machine recovers by itself.
2. Unstable power swing (called out-of-step condition) from which the machine cannot recover. The machine needs to be separated at some point of the grid so that the split systems can feed their individual loads, avoiding large-scale load dislocation.

This out-of-step relay is not necessarily close to the generator, and its location and operation have been discussed in Chapter 4.

8.3. PROBLEMS AND EXERCISES

1. The three-phase stator winding of an alternator can be subjected to the conventional three-phase, L-L-G, L-L and L-G faults at the machine terminals or inside the winding. What sort of relaying is applied to detect the faults? What control action does this relay take?
2. The conventional differential relay (i.e., longitudinal differential relay), when applied to an alternator, does not have a high percent slope (i.e., bias). Explain why.
3. Consider an interturn fault on the same phase. Is is cleared by a longitudinal differential relay? If not, what changes have to be made in the phase windings, and what relay is applied?
4. What is the effect of too much unbalance stator current on a synchronous machine? How is it detected? Is instantaneous action necessary or can the relay be delayed?
5. How close-to-neutral fault detected on one of the alternator phase windings?

Chapter 8

6. Why is the field winding grounded? What is the effect of ground faults at two locations on the field winding? How is the relay connected?
7. Explain what happens to an alternator connected to a grid in the event of
   a. Loss of field excitation
   b. Loss of prime-mover input
   What relays are used?

9
Transformer Protection

9.1. INTRODUCTION

A three-phase transformer, either step-up or step-down, is an integral part of any power system. Small distribution transformers are protected by drop-out fuses, whereas high-capacity transformers require more sophisticated relaying. Such transformers are subjected to three-phase, L-L-G, L-L, L-G and interturn faults, and, the resulting short-circuit current being abnormal, the transformer needs to be disconnected from the remaining healthy system, with the help of reliable relaying, to avoid damage.

9.2. TRANSFORMER CONNECTIONS AND PHASE SHIFT BETWEEN INPUT AND OUTPUT CURRENTS

Figure 9.1 shows the four types of connections (*star-delta*, *delta-star*, *star-star*, and *delta-delta*) for a three-phase transformer. The most common connection for stepping up the generator voltage for transmission purposes is the delta-star, with the HV neutral (star secondary) solidly grounded. Solid grounding ensures minimum insulation requirement on the HV side and, therefore, minimizes transformer cost. Readers can easily find the effect of an A-E fault on healthy phase voltages B-E and C-E in a solidly grounded and ungrounded neutral. For a solidly grounded neutral, V_{B-E} and V_{C-E} remain unaffected as phase to neutral (Fig. 9.2). For an ungrounded neutral, V_{B-E} and V_{C-E} will increase from phase voltage to line-to-line voltage (Fig. 9.2). The healthy phases are stressed to L-L voltage, requiring more insulation, and therefore the transformer costs more. The normal practice is to solidly ground the neutral of the HV side.

To find the phase shift between the input and output currents in the easiest way, connect three single-phase transformers in the desired configuration. Consider first the most commonly used delta-star connections (Fig.

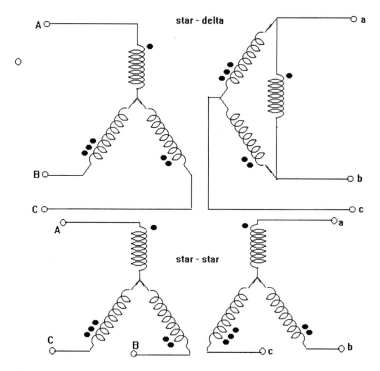

Figure 9.1 Three-phase transformer connections.

9.3). Note carefully the polarity marks on individual single-phase transformers when connecting them (the delta side is the low-voltage side). Assume a 1:1 turns ratio. The neutral point of the star is solidly grounded. Denote the line currents of the three phases on the star side by I_a, I_b, I_c, for phases a, b, c, respectively.

In any transformer the primary and secondary ampere-turns are balanced (neglecting magnetizing current). Therefore the input currents to individual single-phase transformers will be the same as output currents I_a, I_b and I_c. The delta side line input currents are therefore

$I_A = I_a - I_c$ (phase A)

$I_B = I_b - I_a$ (phase B)

$I_C = I_c - I_b$ (phase C)

In Fig. 2c the phasor diagrams of the input and output line currents and voltages show that

Chapter 9

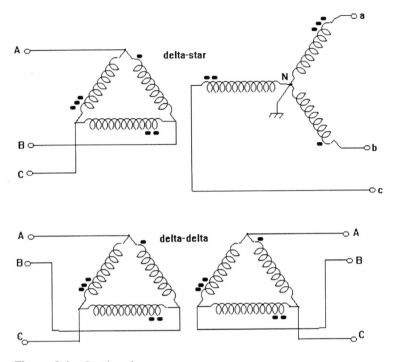

Figure 9.1 Continued

1. The output line currents (star side) lead the input line currents (delta side) by 30°.
2. The delta side line currents are $\sqrt{3}$ times the star side line currents.
3. The star side line voltages are $\sqrt{3}$ times the delta side line voltages.
4. The star side line voltages lead the delta side line voltages by 30°.

Hence,

$$[\sqrt{3}V_L I_L]_{\text{delta}} = [\sqrt{3}V_L I_L]_{\text{star}}$$

VA on delta side = VA on star side

Figure 9.4 summarizes these relations for all types of connections.

9.3. DIFFERENTIAL PROTECTION

Unlike three-phase generators, where the input and output currents are in phase as well as equal, in transformers these currents are different. This was

250   **Transformer Protection**

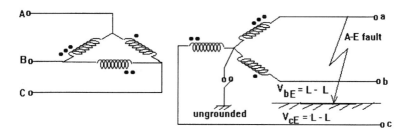

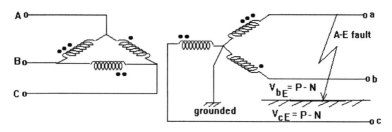

Figure 9.2 Effect of grounding the neutral.

explained in Section II. It is therefore necessary to adjust these currents in phase as well as magnitude before they are fed to the differential relay to ensure that the relay does not operate on normal load flow as well as an external fault. Adjusting the amplitude is called an amplitude check (British usage) or a ratio check (American usage). Adjusting the phase is called a phase check (British) or a phasing check (American).

9.3.1. CT Connections for Amplitude and Phase Balance

We shall take the example of a biased differential relay as applied to the most commonly employed delta-star connections of the power transformer, with the neutral on the star side solidly grounded.

The input/output pilot wire currents fed to the relay are manipulated as following:

    CT turns ratio → for ratio check

    CT connections → for phasing check (delta or star)

Chapter 9

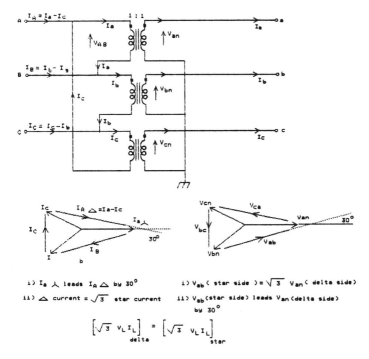

Figure 9.3 Delta-star connections/phase shifts.

The three-phase transformer, as a bank of three single-phase transformers connected in delta on the primary side and in star on the secondary side, is shown in Fig. 9.5. Note carefully the

1. Polarity marks on the power transformers
2. Single-phase transformers with turns ratio of 1:1
3. CT secondaries connected in star on delta side of the power transformer
4. CT secondaries connected in delta on star side of the power transformers
5. Polarity marks on the current transformers

Also note that

1. The secondary (star) side assumed line currents are I_a (phase a), I_b (phase b) and I_c (phase c).

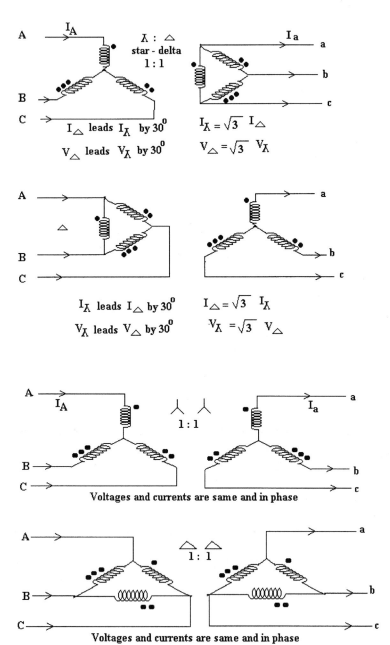

Figure 9.4 Phase shifts for various transformer connections.

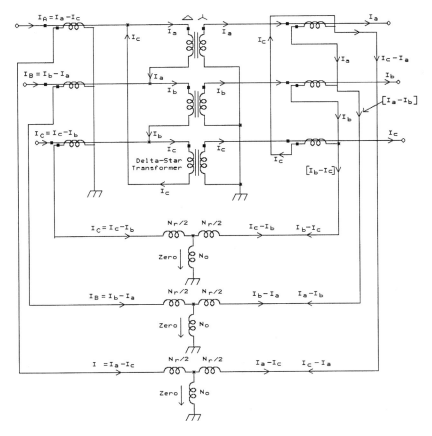

Figure 9.5 Connections for delta-star transformers.

2. The calculated primary line currents (delta side) are

$$I_A = I_a - I_c$$
$$I_B = I_b - I_a$$
$$I_C = I_c - I_b$$

3. From the phasor diagram of the currents (Fig. 9.5):
   a. The delta side line currents are $\sqrt{3}$ times star side line currents.
   b. The delta line currents lag the star side line currents by 30°.

To eliminate the magnitude and phase differences in input/output currents before they are fed to the relay, one must connect the CTs as follows:

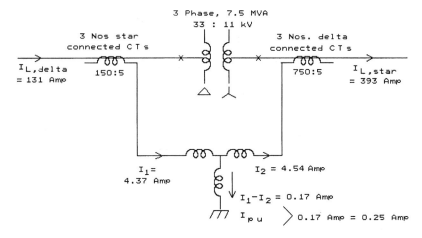

Figure 9.6 Numerical example.

1. The CT secondaries must be connected in star on the delta side of the power transformer.
2. The CT secondaries must be connected in delta on the star side of the power transformer.
3. All CT ratios should be properly chosen for amplitude balance.
4. The relay must not operate on a normal load or an external fault.

Similarly the CT secondary connections for different configurations of power transformers can be evaluated and is left to the reader.

9.3.2. Pickup and Percentage Slope

Consider the following example. A three-phase, delta-star, step-down transformer rated 7.5 MVA, 33 to 11 kV, is to be protected by a biased differential relay. Draw the complete connections and calculate the CT ratios. The CT ratios may be rounded to the nearest values.

solution

$$I_{L,\text{star}} = 393 \text{ A} = 7500 \text{ KVA}/\sqrt{3} \times 11 \text{ KV}$$

CT ratio = $(\sqrt{3}\ 393):5$ A (delta side)

= 750:5 A (delta connected)

Similarly,

$I_{L,\text{delta}} = 131$ A

CT ratio = 150:5 A (star connected)

Chapter 9

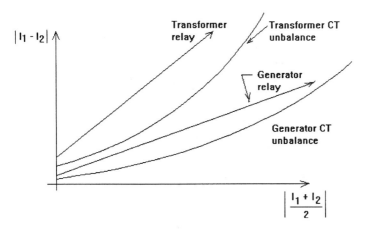

i) Transformer relay ; high pick-up ;high slope

ii) Generator relay , low pick-up ; low slope

Figure 9.7 Percentage slope for generator and transformer differential relaying.

The pilot wire currents or currents fed to differential relay are

$$I_1 = \frac{131 \times 5}{150} = 4.37 \text{ A} \quad (33\text{-kV side})$$

$$I_2 = \frac{\sqrt{3} \times 393 \times 5}{750} = 4.54 \text{ A} \quad (11\text{-kV side})$$

$$I_1 - I_2 = 0.17 \text{ A ac} \quad (\text{CT unbalance current})$$

$$I_{pu} = 5\% \text{ of } 5 \text{ A} = .25 \text{ A} > 0.17 \text{ A}$$

Figure 9.6 shows the complete circuit.

Unlike generators, where the input/output CTs are identical and have the same ratio, the CTs for transformer protection have different ratios and, hence, different saturation levels. The CT unbalance current, therefore, is much higher in transformer relaying compared to generator relaying. This is shown in Fig. 9.7. Note that the pickup and percentage slope are also much higher for transformer relaying than for generator relaying.

9.3.3. Example of L-G Internal and External Faults

Figure 9.8 shows A-E internal and external faults on the star side of the delta-star three-phase transformer. Note that a fault before the CT is an

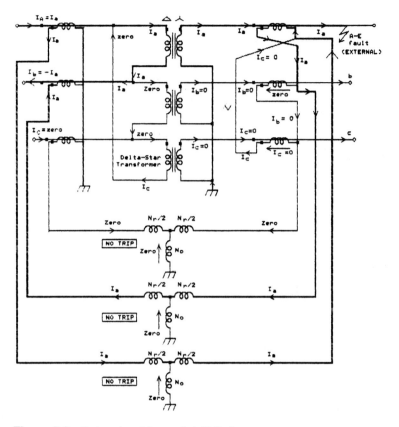

Figure 9.8 External and internal A-E faults.

internal fault, whereas a fault outside the CT position is an external fault. The following conclusions can be drawn:

1. A single L-G fault on the star side appears as a L-L fault on the delta side.
2. For an external fault none of the differential relays operate.
3. For an internal L-G fault two relays operate.
4. It is possible to some extent to diagnose the type of fault in the transformer from the target indication of the relays that have operated.

Chapter 9

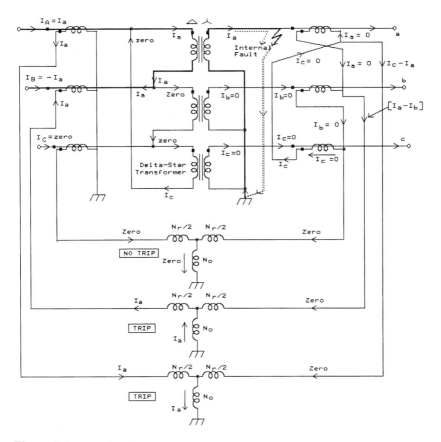

Figure 9.8 Continued.

9.4. RESTRICTED EARTH-FAULT RELAY

In Figure 9.9 notice that a L-G fault close to the neutral results in a very small fault current, and the conventional differential relay may fail to operate. Therefore, we need an additional earth-fault relay sensitive to ground faults in the transformer winding close to the neutral.

The commonly used generator/step-up transformer configuration is shown in Fig. 9.10, along with an instantaneous overcurrent relay in the neutral of the star connections of the secondary transformer winding in the

258   **Transformer Protection**

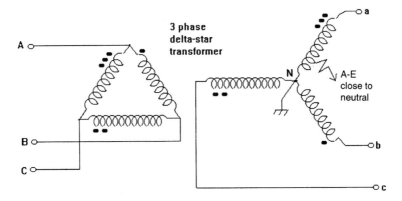

i) Fault close to neutral ; fault current small

ii) No differential relay would operate

Figure 9.9   L-G fault close to neutral.

event of L-G faults. This connection is not correct since an external or through fault on the bus or HV line will operate the relay. This necessitates restricting the reach of the transformer ground-fault relay so that it does not operate on external faults.

The correct connections are shown in Fig. 9.11, where the three line CTs and another CT in the neutral are paralleled and the overcurrent relay is connected across this parallel combination. The internal fault is that fault which takes place before the line CTs, and the external fault is one which

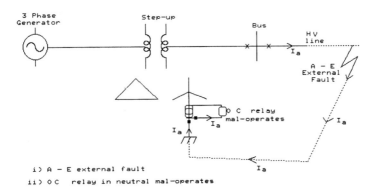

Figure 9.10   Ground-fault relay operates on through fault.

Chapter 9

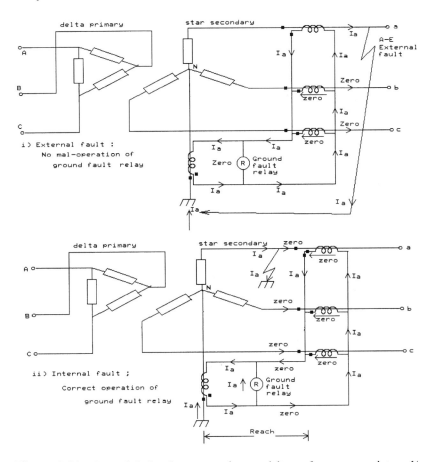

Figure 9.11 Ground-fault relay connections and its performance on internal/external faults.

takes place beyond the line CTs. The relay performance for internal and external faults is also shown in Fig. 9.11.

9.5. MAGNETIZING INRUSH CURRENT

When the power transformer is energized on the primary side, with the secondary open-circuited, the current drawn by the primary is the magnetizing current, which, in steady state, is about 5% (or even less) of the full-load current. This current, however, in the first few cycles (called *inrush* current) can be very large and even exceed the full-load current. This is explained in subsequent sections.

9.5.1. Steady-State and Transient Magnetizing Currents and Their Effect on Differential Relays

Consider for simplicity a single-phase transformer protected by a differential relay (Fig. 9.12). If we consider only the steady state, the current in the spill circuit or the relay is approximately 5% or less of the transformer full-load current. Since no fault has occurred, any possible relay malfunction can be averted by choosing the pickup value to be greater than 5% full-load current. This choice, however, is ineffective since, recall, that the transient inrush current, in the first few cycles, can be as large as the full-load current. Hence, the pickup value must be more than the full-load current. However, the minimum internal fault current for which the relay is required to operate is then too large—clearly an impossible solution. The correct solution is then to block relay operation on transient magnetizing inrush current and choose a small pickup value in the event of light internal faults.

9.5.2. Effect of Residual Flux, Instant of Energization, and Nonlinear B-H Curve

Consider a single-phase transformer (Fig. 9.13) where the secondary is open-circuited and the primary is energized by a sinusoidal voltage at the instant

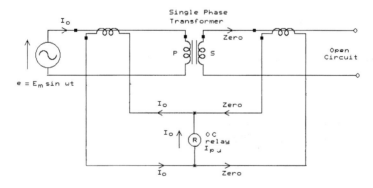

Figure 9.12 Malfunction due to magnetizing current.

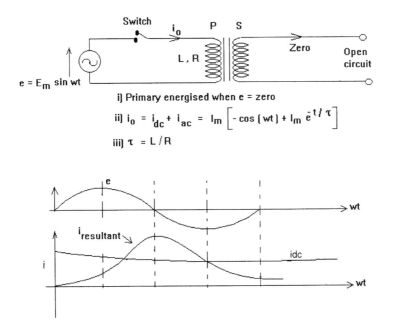

Figure 9.13 Dc offset in magnetizing current.

the voltage passes through its zero value. The primary has its own self-inductance L and resistance R. In Chapter 4 it was proved that

$$i = I_m[\underbrace{-\cos wt}_{\substack{\text{ac}\\\text{component}}} + \underbrace{I_m e^{-t/T}}_{\substack{\text{dc}\\\text{component}}}]$$

where $T = L/R$. The input voltage and current waveforms are shown in Fig. 9.13. Clearly, in the first few cycles the appropriate rms value of the current is

$$I_{\text{magnetizing}} = \sqrt{I_{ac}^2 + I_{dc}^2}$$

To prevent the differential relay from operating, we must increase the pickup value, but this is not the end of the problem.

We shall now see the effect of a nonlinear B-H curve and the residual flux on the magnetizing inrush current. Refer to Fig. 9.14. Again a single-phase transformer is switched into the sinusoidal voltage source at its zero value. Also assume that prior to switching there is a residual flux ϕ_r, which, under worst conditions, could be $\phi_{\text{max,ac}}$, which is the peak flux value for the input sinusoidal voltage $e = E_m \sin wt$. According to the figure:

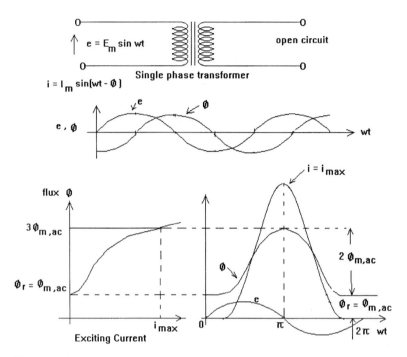

Figure 9.14 Effect of nonlinear B-H curve and residual flux on magnetizing current.

1. $e = E_m \sin wt$, the input voltage waveform.
2. The instant of switching occurs at the voltage zero crossing.
3. The flux excursion from residual flux is $\phi_r = \phi_{max,ac}$.
4. The maximum excursion of the flux is $3\phi_{max,ac}$.
5. The magnetic core, designed for $\phi_{saturation} = \phi_{max,ac}$, must produce three times this value, hence the core will saturate.
6. The magnetizing inrush current will have a dc offset and numerous harmonics. This waveform is also shown in the figure.
7. Fourier analysis gives the following dc and harmonic contents in a typical magnetizing inrush current:
   a. dc = 55%
   b. Fundamental = 100%
   c. 2nd harmonic = 63%
   d. 3rd harmonic = 27%
   e. 4th harmonic = 5%
   f. 5th harmonic = 4%

Chapter 9

g. 6th harmonic = 3.7%
h. 7th harmonic = 2.4%

Various suggestions have been made as to how to restrain the operation of the differential relay for such high values of inrush current while keeping the pickup value low enough so that low-level internal faults can be effectively cleared.

9.5.3. Harmonic Restrained Differential Relay

Here are two suggestions for avoiding malfunction of the differential relay on magnetizing inrush currents:

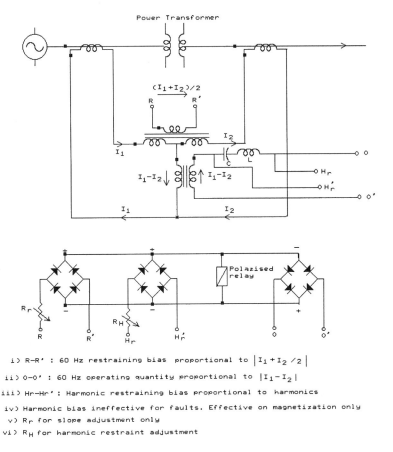

i) R–R' : 60 Hz restraining bias proportional to $|I_1 + I_2 / 2|$
ii) O–O' : 60 Hz operating quantity proportional to $|I_1 - I_2|$
iii) Hr–Hr' : Harmonic restraining bias proportional to harmonics
iv) Harmonic bias ineffective for faults. Effective on magnetization only
v) R_r for slope adjustment only
vi) R_H for harmonic restraint adjustment

Figure 9.15 Harmonic restrained biased differential relay.

1. Since the second harmonic is dominant, filter it out and apply an additional restraint to the differential relay.
2. Attempt to remove all harmonics, including the dc offset, and apply it as an additional restraint.

These relays are called harmonic restrained differential relays. One static version of this, using full-wave diode bridges and a filter, is shown in Fig. 9.15.

9.6. OVERFLUXING PROTECTION

Faraday's law of induction gives the following relation for any transformer excited by a sinusoidal voltage:

$$e = N \frac{d\phi}{dt}$$

Substituting $e = E_m \sin wt$ and integrating for flux gives the famous relation

$$E = 4.44 f N \phi$$

Thus, the rms value of flux is proportional to E/f, or it is directly proportional to the applied ac voltage E and inversely proportional to frequency f of the ac voltage.

The overfluxing relay is shown in Fig. 9.16, where the ac voltage from a suitably placed PT is integrated; if this integrated value exceeds a value corresponding to 110% of the rated value, the overvoltage relay, acting as an overfluxing relay, gives an alarm.

The analytical relations are

$$e = N \frac{d(\text{flux})}{dt}$$

or

$$\text{Flux} = \frac{1}{N} \int e \, dt$$

The circuit consists of a very large resistance R in series with a capacitor C. The analytical relations are

$$i = \frac{e_i}{R}$$

Chapter 9

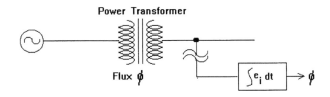

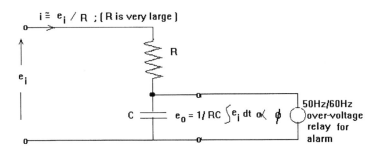

Figure 9.16 Overfluxing relay.

Now

$$e_0 = \frac{1}{C} \int i \, dt = \frac{1}{RC} \int e_i \, dt$$

If the rms value of e_0 exceeds the set value, an alarm is given indicating either overvoltage or underfrequency.

9.7. BUCHHOLZ RELAY/SUDDEN PRESSURE RELAY FOR INCIPIENT FAULTS

Due to insulation aging or water vapor in the transformer oil, the transformer develops a fault over a period of time. Such slowly developing faults are called *incipient* faults. The protection employed for this fault should give an alarm so that corrective action can be taken before the fault worsens and damages the transformer entirely. In case the fault tends to be serious, the circuit breakers can be tripped and the transformer isolated from the healthy system. Thus, the protection is a dual type: (a) for incipient faults an alarm is given and (b) for violent faults the CBs are tripped.

The great majority of high-power transformers are oil-cooled with a conservator and are subjected to incipient and violent faults. The protection employed is the Buchholz relay. Figure 9.17 shows one version of it. The

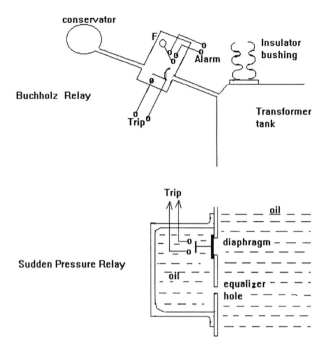

Figure 9.17 Buchholz relay/sudden pressure release relay.

basic principle is based on how quickly the oil breaks down into its various elements (methane gas, etc.) and develops pressure in the transformer oil tank while lowering of the oil level in the conservator over time.

For slowly developing faults, oil breakdown leads to a drop in oil level in the conservator and the float tilts, closing the mercury switch. The mercury switch is wired to give an alarm.

For a violent internal fault, the oil surges from the tank to the pivoting baffle plate of the relay. The plate tilts another mercury switch, which is wired to trip the circuit breakers. Such protection also serves as backup to the standard differential scheme.

In transformers with no conservator tank, it is difficult to install a Buchholz relay. In such cases a sudden pressure relay is built into the top side of the transformer tank itself and is shown in Fig. 9.17.

Under steady-state conditions the small equalizer hole equalizes the pressure between the relay tank (i.e., small one) and the main transformer oil tank. For incipient faults there is a differential pressure against the diaphragm, thereby closing the relay contacts. The relay contacts are invariably wired to trip the circuit breakers.

Chapter 9

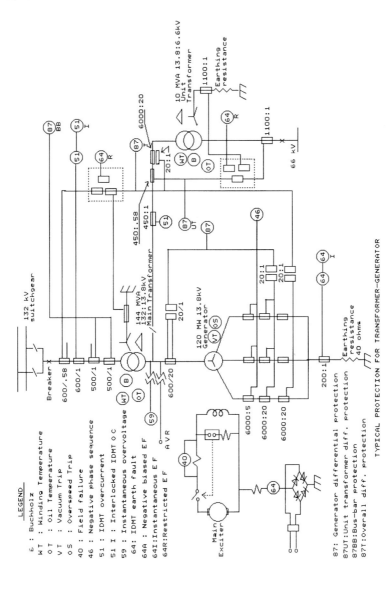

Figure 9.18 Unit protection.

The appropriate settings are 20.0 gcm^{-2} for differential gas pressure and 5.0 gcm^{-2} sec^{-1} for rate of pressure rise.

9.8. GENERATOR/TRANSFORMER UNIT PROTECTION

In most high-capacity turboalternator sets, there is, in addition to the generator and step-up transformer, a unit transformer (usually delta-star) to supply the auxiliaries. The total protection for this combination is given in Fig. 9.18.

9.9. PROBLEMS AND EXERCISES

1. For a single-phase transformer, prove that an interturn fault will be detected by the differential relay.
2. Why are ratio and phasing checks necessary for protecting a three-phase transformer? In a delta-star (star solidly grounded) transformer, what is the rule-of-thumb for connecting CT secondaries on a delta-star power transformer?
3. a. Why is an earth-fault relay used with a transformer?
   b. Why is the reach of this ground-fault relay restricted?
4. Why is the star side (HV side) solidly grounded for a HV step-up transformer?
5. Explain the construction of the Buchholz/sudden pressure relay.
6. For each of the following conditions, explain why the magnetizing current of the transformer tends to increase:
   a. Instant of switching the transformer on input voltage waveform
   b. Nonlinear B-H characteristics of the transformer core
   c. Residual flux

10
Bus Bar Protection

10.1. INTRODUCTION

Bus bars are an integral part of any power system. A bus bar has incoming and outgoing lines terminated on it with appropriate circuit breakers. Bus bar relaying should generate a primary protective zone, overlapping the protective zones of all adjacent lines (Fig. 10.1). Thus, the bus bar is a unit type of protection, incapable of giving any backup protection. The positions of internal and external faults are also shown. For internal faults the relaying should operate and trip all circuit breakers enclosed in the protective zone. For external faults the relaying should not operate.

Since the input and output ends are physically close, we use the differential scheme to protect bus bars. The difference between the conventional differential scheme and the bus bar differential scheme is that a stabilizing resistance is inserted in the spill circuit to stabilize the relay (i.e., to avoid malfunction) on external faults. For this reason the scheme is called a *high-impedance differential scheme*. "High impedance" refers to the impedance of the spill circuit (due to its stabilizing resistance).

10.1.1. Current Transformer Ratio

The normal choice of CT ratio is

$$\text{CT ratio} = \frac{I_{L,\max}}{5.0}$$

This rule does not hold for bus bar relaying, as shown in Fig. 10.2. Let lines A, B and C be terminated on the bus. Lines A and B supply load currents of 100 and 200 A, respectively. Therefore, line C delivers 300 A to the load.

The bus bar relaying is now as follows. The current transformers on all three lines are paralleled (with correct polarity marks), and this combi-

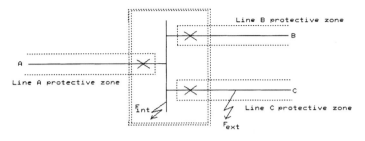

BUS – BAR PROTECTIVE ZONE
OVERLAPPING ON ALL ADJACENT PROTECTION

Figure 10.1 Bus bar protection zone.

nation is connected across an instantaneous overcurrent relay. Let the CT ratios (which are wrong) be

$$\text{CT A} \quad \frac{I_{L,\max,\text{line A}}}{5.0} = 100:5.0$$

$$\text{CT B} \quad \frac{I_{L,\max,\text{line B}}}{5.0} = 200:5.0$$

$$\text{CT C} \quad \frac{I_{L,\max,\text{line C}}}{5.0} = 300:5.0$$

Under normal power flow the current through the OC relay in the spill circuit must be zero. We do not want the relay to operate, since it will result in unnecessary tripping of all CBs when the bus bar is healthy.

Figure 10.2 shows that the spill current is 5.0 A, resulting in unnecessary relay operation during normal power flow. Therefore, in bus bar relaying, all CT ratios must be equal to

$$\text{CT ratio} = \frac{\text{Current in that line where load is maximum}}{5.0}$$

$$= 300:50 \quad \text{(in Fig. 10.2)}$$

Figure 10.3 clearly shows that with all CT ratios equal the OC relay current is zero.

The CTs chosen with these considerations are satisfactory only if the CT load power factor is matched with the power factor (X/R ratio) of modern high-voltage lines. If the CT load is resistive, the CT volt-amperes must be increased by $(1.0 + X/R)$ times the steady-state rated VA to avoid saturating the CTs.

Chapter 10

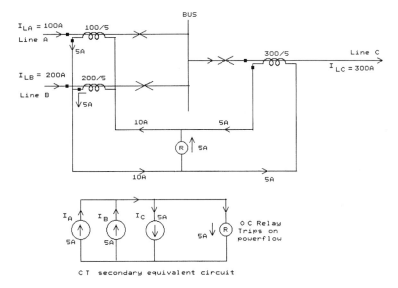

Figure 10.2 Incorrect CT ratio.

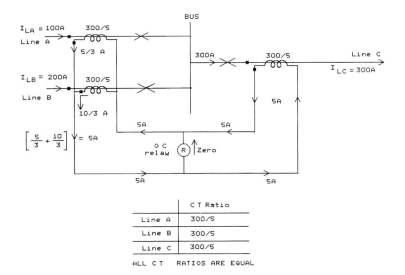

Figure 10.3 Correct CT ratios/no operation on load flow.

10.1.2. External Fault

Consider Fig. 10.4, where three lines are terminated on the bus bar. Assume that there are sources of power behind all three lines. The normal power flow prior to any fault is

    line A   100 A
    line B   200 A
    line C   300 A

As explained previously, all three CT ratios must be to 300:5.0 A.

Now consider an external fault on line C (i.e., beyond CT position). The assumed fault currents are

    line A   1000 A
    line B   2000 A
    line C   1500 A

    Total fault current = 4500 A

The CT currents are

    CT A   1000 A
    CT B   2000 A
    CT C   3000 A (not 4500 A)

Figure 10.4 shows the equivalent circuit of the differential scheme. All CTs are assumed to be ideal and not saturated. The current in the spill circuit or OC relay is now zero. Since the fault is external, the relay does not operate.

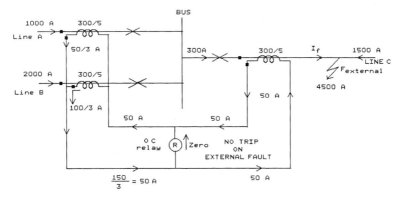

Figure 10.4 External fault/ideal current transformers.

Chapter 10

Intuitively, however, CT C, carrying a maximum fault current of 3000 A, is likely to be saturated. The effect of this saturation will be dealt with in the following sections.

10.1.3. Internal Fault

Figure 10.5 is the same figure as Fig. 10.4, except that the fault is located in line C, so it is an internal fault. Note that the currents in the CT primary windings are the same as for the external fault just discussed.

From CT equivalent circuits of Fig. 10.5, the fault current in the OC relay is now the total fault current of 75 A (15 times the rated current of 5.0 A), which causes the relay to operate and isolate the bus from all three lines. This operation is correct. Note that CT saturation, if any, has been neglected.

10.2. HIGH-IMPEDANCE DIFFERENTIAL SCHEME

Up to now the bus bar relay operation appeared satisfactory, but it is not, due to unequal saturation of the CTs. Note that each CT is carrying a different primary current, and it is obvious that the CT carrying maximum

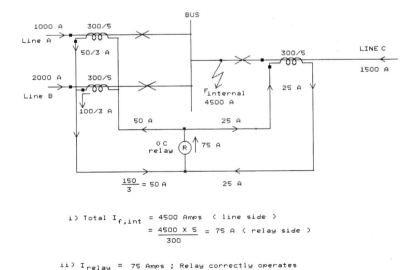

Figure 10.5 Internal fault/ideal current transformers.

primary current is likely to be saturated. At this stage it is necessary to recall the magnetizing impedances of unsaturated and fully saturated CTs.

10.2.1. External Faults with CT Equivalent Circuit

The equivalent circuits of unsaturated and fully saturated CTs are shown in Fig. 10.6. For brevity assume that, for an unsaturated CT, the magnetizing impedance $Z_0 = \infty$. Therefore $I_0 = 0$ and $I_S = I_p/N$. For a fully saturated CT, $Z_0 = 0$. Therefore, $I_0 = I_p/N$ and $I_S = 0$.

Figure 10.7 shows three line current transformers paralleled after the CT secondary resistance and the lead resistance of the relay pilot wires. (Note that the CTs are paralleled in the control room, where the relay is located, and not in the switchyard, where the CTs are located. The CTs are usually placed in the bushings of the circuit breaker.) The instantaneous overcurrent relay along with a stabilizing resistance are wired in parallel with the three current transformers.

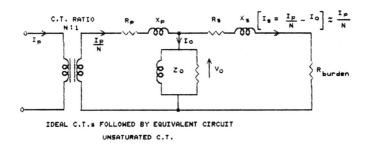

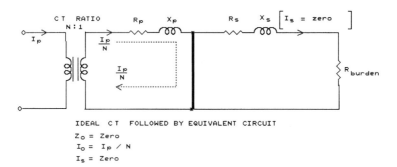

Figure 10.6 Equivalent circuits of saturated and unsaturated current transformers.

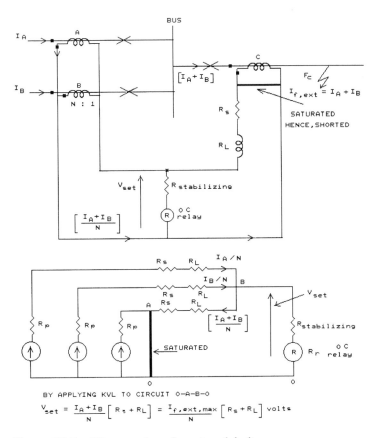

Figure 10.7 CT connections for external fault.

Note that CT C, which is on the faulted feeder, is assumed to be completely saturated and, therefore, short-circuited. The external fault current flows through the parallel combination of the CT secondary resistance and the relay circuit. Since the CT secondary resistance is very small, it is reasonable to assume that most of the fault current flows through it. Therefore, the voltage across the relay circuit, V_{set}, called the *setting voltage* is

$$V_{\text{set}} = \frac{I_{\text{f,ext,max}}}{N}(R_s + R_L)$$

where $I_{f,ext,max}$ = maximum through-fault current contributed by all lines terminated on the bus
R_s = CT secondary resistance, assumed to be equal for all CTs
R_L = lead resistance or relay pilot wire resistance; CT secondary leakage reactance is neglected

It is absolutely necessary that, for all CTs, $V_{knee} > V_{set}$, to avoid saturation of CTs on unfaulted feeders.

Let the spill circuit OC relay have current I_{pu}, and let the total resistance of the spill circuit be $R_{stabilizing} + R_{relay}$. To ensure that the relay not operate on external faults, the following relation exists between various relay parameters and settings:

$$I_{pu} \geq \frac{V_{setting}}{R_{stabilizing} + R_{relay}}$$

$$I_{pu} \geq \frac{I_{f,ext}(R_{secondary} + R_{lead})}{N(R_{stabilizing} + R_{relay})}$$

As explained in the following section, the selected I_{pu} value also decides the minimum value of internal fault current for which the relay operates satisfactorily. Thus, knowing a priori I_{pu}, the value of $R_{stabilizing}$ can be calculated to avoid any relay malfunction. Since the stabilizing resistance increases the total impedance in the spill circuit or relay circuit, this scheme is called a *high-impedance bus bar differential scheme*.

10.2.2. Internal Faults with CT Equivalent Circuit

Figure 10.8 shows the current transformer secondary connections along with the relay circuit. As pointed out earlier, for internal faults the line currents are relatively small, unlike through faults, where the total current flows in the faulted feeder.

No current transformer is assumed to be saturated, and each one absorbs a magnetizing current I_0 at an excitation voltage of V_{set}, as read from the CT secondary excitation characteristics. Here,

$$V_{set} = \frac{I_{f,ext,max}(R_s + R_L)}{N} < V_{knee}$$

N = CT ratio

If q is the number of lines (or no. of CTs) terminated on the bus bar, we have

Total $I_0 = qI_0$

Chapter 10

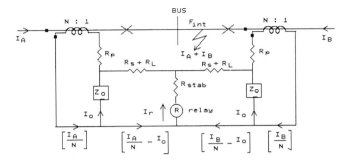

$$I_r = \frac{I_A + I_0}{N} - 2I_0 \quad ; \text{(for simplicity only two buses assumed)}$$

$$= \frac{I_{f,int}}{N} - 2I_0$$

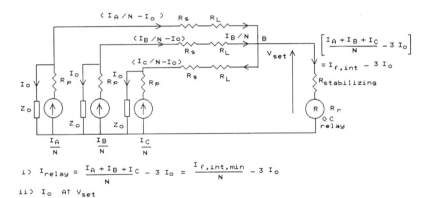

i) $I_{relay} = \frac{I_A + I_B + I_C}{N} - 3I_0 = \frac{I_{f,int,min}}{N} - 3I_0$

ii) I_0 AT V_{set}

Figure 10.8 CT equivalent circuit for internal faults.

The current in the relay is now

$$I_{relay} = \frac{I_{f,int,min}}{N} - qI_0$$

The relay operates if

$$I_{pu} \leq I_{relay}$$

or

$$I_{f,int,min} \geq N(I_{pu} + qI_0)$$

Note that the smaller the CT's magnetizing current, the smaller is the internal fault current for which relay operates satisfactorily. This necessitates good quality CTs.

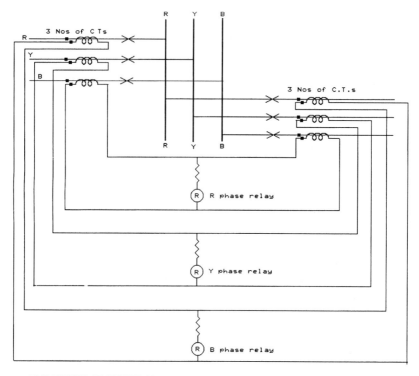

i) 3 NUMBERS OF DIFFERENTIAL RELAYS FOR ALL FAULTS (3 PH , L-L-G , L-L ,L-G)
ii) RELAYS WIRED TO TRIP ALL INCOMING / OUTGOING BREAKERS

Figure 10.9 Three-phase bus bar differential scheme.

So far, we have used single-line diagrams of the bus bars. In reality the bus bars are three-phase and are likely to be subjected to three-phase, L-L-G, L-L and L-G faults, a total of 10 faults. To detect all them a three-phase scheme is required, as shown in Fig. 10.9.

10.3. STABILITY RATIO

Recall that external and internal fault currents occasionally vary due to the type of fault (the L-G fault normally gives the minimum current, whereas the three-phase fault gives maximum current) and the variation in source impedance (maximum source impedance results in minimum fault current, and minimum source impedance results in maximum fault current). The stability ratio, S, for this scheme is defined as

Chapter 10

$$S = \frac{I_{f,ext,max}}{I_{f,int,min}}$$

= as large as design permits

> 30

where for $I_{f,ext,max}$ the relay does not malfunction, for $I_{f,int,min}$ the relay operates satisfactorily.

10.4. CHECK FEATURE AND SUPERVISORY CIRCUIT

Two lines of defense are invariably applied for all EHV bus bars. In the various bus bar arrangements, it may happen that some lines are terminated on one bus bar and the rest on the other bus bar. The check feature is more or less identical to the conventional high-impedance bus bar differential scheme, except that it supervises all the incoming and outgoing lines, regardless of their termination on different bus bars (in other words, the check feature cannot discriminate which bus is involved in the fault) and separate current transformers are used.

Figure 10.10 shows a single line diagram of the differential scheme for bus A, the differential scheme for bus B and an overall check feature incorporating both buses.

On all important bus bar installations, it is absolutely necessary to check the continuity of CT secondaries and the relay pilot wires so that the differential scheme is alert and ready to deal with internal faults at any time (Fig. 10.11).

Under various conditions, the spill currents are

$$I_{relay,power\,flow} = (q - 1)I_0$$

$$I_{relay,power\,flow,one\,CT\,out} = \frac{I_L}{N} - (q - 1)I_0$$

Thus, a loss of CT or pilot wire appears as an internal fault, and the differential relay may malfunction. Therefore, if a CT or relay pilot wire is lost, a *supervisory relay* must detect this condition, give an alarm and block possible malfunctioning of the bus differential scheme (note that there is no internal fault on the bus bar) by short-circuiting it by a hand-reset relay. This is shown in Fig. 10.12 on a single-phase basis. How to calculate the supervisory relay setting is discussed in the numerical example.

The effective setting of the supervisory relay is determined by the magnetizing current drawn by the current transformers and by the current shunted by the high-impedance bus differential scheme, and should be as

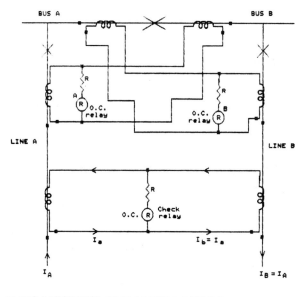

i) ZONE A DIFFERENTIAL RELAY : (LINE A + BUS B)
ii) ZONE B DIFFERENTIAL RELAY : (LINE B + BUS A)
iii) CHECK RELAY : (LINE A + LINE B)

Figure 10.10 Check feature.

small as possible. It is normal practice to keep the setting at 25.0 A or 10% of $I_{L,\min}$, whichever is greater. It has been reported that if $V_{\text{set}} = 100$ V for the high-impedance differential scheme with $I_{\text{pu}} = 0.8$ A, the setting of the supervisory relay is as low as $I_{\text{pu,supervisory}} = 0.00123$ A and $V_{\text{set,supervisory}} = 2.0$ V.

10.5. NUMERICAL EXAMPLE

Consider a 132-kV bus bar scheme (Fig. 10.13) where two incoming lines and four outgoing lines are terminated. The system is solidly earthed, and the switchgear rating is 3500 MVA at 132 kV. The parameters are

Maximum full-load current in one line	500 A
All CT ratios	500:1.0 A
R_s (CT secondary resistance)	0.7 ohm
$R_{\text{lead wire}}$	2.0 ohms
Relay load	1.0 VA

Chapter 10

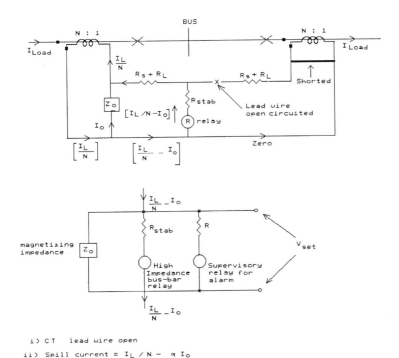

i) CT lead wire open
ii) Spill current = $I_L/N - q\, I_o$
iii) q = number of lines terminated on bus and = 2 in the above figure

Figure 10.11 Relay current on normal power flow with one CT lost.

Load R	VA/I^2 ohms
CT magnetizing current I_0 up to 120 V	0.28 mA/V (linear)
CT saturation voltage V_{knee}	>120 V
Maximum through-fault current	$\dfrac{3500 \text{ MVA}}{\sqrt{3}\ 132 \text{ kV}}$
	15,300 A

The relay constraints are

1. The relay must not operate for $I_{f,ext,max}$ = 15,300 A.
2. The relay must operate for $I_{f,int,min}$ = 500 A (the load current).
3. The desired stability ratio ≥ 15,300/500 = 30.6

Calculate the following:

1. Relay voltage setting to avoid malfunction for through-fault current of 15,300 A

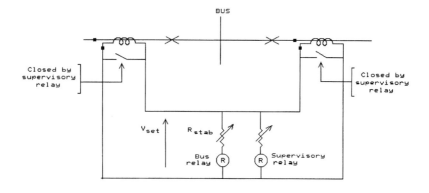

i) V_{set} for supervisory relay $\ll V_{set}$ for bus-bar scheme

ii) CT open-circuited on any feeder ; supervisory relay operates

iii) Supervisory relay takes the protection out of service
(by shorting CT secondaries) and gives alarm

Figure 10.12 Supervisory relay.

2. Relay current setting for correct operation for a minimum internal fault current of 500 A
3. $R_{stabilizing}$
4. S
5. Supervisory relay setting ≤ 25.0 A

Solution
1.

$$V_{set} = \frac{I_{f,ext,max}(R_0 + R_L)}{N}$$

$$= \frac{15{,}300(2.0 + 0.7)}{500}$$

$= 82.6$ V (well below V_{knee})

$= 100.0$ V (assumed as safety margin)

2.

At 100.0 V $I_0 = 0.28 \times 100$ mA

$= 28$ mA $= 0.028$ A (for each CT)

Chapter 10

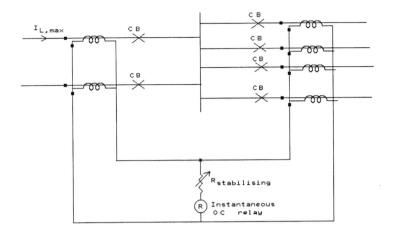

  i) SWITCHGEAR RATING : 132 kV, 3500 MVA
 ii) $I_{L,max}$: 500 Amp
iii) CT SECONDARY RESISTANCE = 0.7 OHM
 iv) LEAD WIRE RESISTANCE = 2.0 OHMS
  v) CT MAGNETISING CURRENT = 0.28 mA / Volt UPTO 120 Volts
 vi) V_{knee} FOR CT = 120 Volts

Figure 10.13 Numerical example on bus bar.

For six CTs $I_0 = 6.0 \times 0.028$ A
              $= 0.168$ A

Assume a 1.0-A relay with a rating of 1.0 VA. Therefore, the load in ohms is

$$R_L = \frac{VA}{I^2} = 1 \text{ ohm}$$

The relay loads at various relay pickup values are

Relay I_{pu}	1.0	0.4	0.3	0.2	0.15	0.1
R_L (ohms)	1.0	6.28	11.0	25.0	44.0	100

$$I_{f,int,min} = N(I_{pu} + qI_0)$$
$$500 = 500(I_{pu} + 0.168)$$
$$I_{pu} = 0.832 \text{ A}$$

We round off I_{pu} to 0.9 A. Then
$$I_{f,int,min} = 500(0.9 + 0.168)$$
$$= 534 \text{ A}$$

Thus, a current setting of 0.9 A is satisfactory and clears internal faults of 534 A and above.

3.
$$I_{pu} \geq \frac{I_{f,ext}(R_{secondary} + R_{lead})}{N(R_{stabilizing} + R_{relay})}$$

$$I_{f,ext,max} = \frac{I_{pu}N(R_{stabilizing} + R_{relay})}{R_{secondary} + R_{lead}}$$

$$15{,}300 \text{ A} = \frac{0.9 \times 500(R_{stabilizing} + R_{relay})}{2.0 + 0.7}$$

Therefore,
$$R_{stabilizing} + R_{relay} \geq \frac{15{,}300 \times 2.7}{0.9 \times 500}$$
$$\geq 92.0 \text{ ohms}$$

and
$$R_{stabilizing} \geq 92.0 - R_{relay}$$
$$\geq 92.0 - 1.0$$
$$= 91.0 \text{ ohms}$$

4.
$$S = \frac{I_{f,ext,max}}{I_{f,int,min}}$$
$$= \frac{15{,}300 \text{ A}}{534 \text{ A}}$$
$$\approx 29$$

5. For setting the supervisory relay see Fig. 10.14, which shows the equivalent circuit of the supervisory relay, consisting of a parallel combination of (a) a low-set supervisory relay, (b) a high-impedance bus differ-

Chapter 10

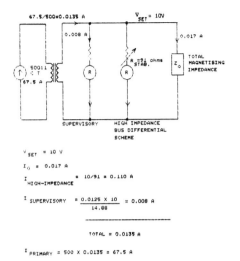

$V_{SET} = 10$ V
$I_0 = 0.017$ A
$I_{HIGH-IMPEDANCE} = 10/91 = 0.110$ A
$I_{SUPERVISORY} = \dfrac{0.0125 \times 10}{14.88} = 0.008$ A

TOTAL = 0.0135 A

$I_{PRIMARY} = 500 \times 0.0135 = 67.5$ A

Figure 10.14 Equivalent circuit of supervisory relay.

ential scheme and (c) total magnetizing impedance. Hence, the primary operating current, which is lost, due to a loss of a CT is

CT ratio × [supervisory relay setting
+ current shunted by high-impedance bus scheme
+ total magnetizing current]

Let

$I_{p,\text{line side}} = 25.0$ A   (minimum load current lost due to loss of a CT)

$I_{\text{relay side}} = \dfrac{25 \times 1}{500} = 0.05$ A

q = number of circuits = 6.0

$V_{\text{set,supervisory}} = 15$ V

$I_{\text{pu,supervisory}} = 0.0125$ A   (assumed)

$R_{\text{supervisory relay}} = 1200$ ohms

Let half the current be absorbed as magnetizing current. Then for the six CTs,

$$I_0 = 0.5 I_{\text{relay side}}$$
$$= 0.5 \times 0.05 \text{ A}$$
$$= 0.025 \text{ A}$$

For an individual CT,

$$I_0 = \frac{0.025}{6} = 0.00416 \text{ A}$$

Since the magnetizing current is 0.28 mA/V,

$$V_{\text{set}} = \frac{0.00416 \times 1000}{0.28} = 14.88 \text{ V}$$

To be on the safe side choose $V_{\text{set}} = 10$ V (instead of 14.88 V). Then the total magnetizing current is

$$I_0 = \frac{0.28 \times 10 \times 6}{1000} = 0.017 \text{ A} \qquad \text{(instead of 0.025 A)}$$

and

$$I_p = 500 \left(0.017 + \frac{0.0125 \times 10}{14.88} + \frac{10}{91} \right)$$
$$= 500 \times 0.135$$
$$= 67.5 \text{ A} \qquad \text{(assumed value before design is 25.0)}$$

For convenience we list the answers here:

1. Voltage setting   100.0 V
2. Current setting   0.9 A
3. Stabilizing resistance   91 ohms
4. Stability ratio   29
5. Supervisory relay setting   67.5 A on line side

10.6. PROBLEMS AND EXERCISES

1. Explain why the current transformer ratios have to be identical for the bus bar differential scheme.
2. Prove that for current transformers
   a. $I_s = I_p/N - I_0$ for unsaturated CT
   b. $I_s = 0$ for saturated CT
3. Derive equations for the following:
   a. Maximum fault current beyond which the relay malfunctions

Chapter 10

      b. Minimum internal fault current for which the relay operates correctly.
4. Define stability ratio. Why should it be as high as the design permits?
5. What is the purpose of the supervisory relay? Draw the three parallel branches that constitute the supervisory scheme.
6. Why is the check feature necessary, say in a sectionalized single bus bar arrangement? Show the current transformer positions for the two sections of the bus and the overall bus.

11
Test Procedures, Benches, and Maintenance Schedules

11.1. INTRODUCTION

All protective devices must remain constantly alert for faults and take appropriate action. Faults may take place, perhaps, only once in 10 years, but the relaying has to remain operational and ready. Hence routine testing (both steady-state and dynamic) is necessary. In steady-state testing the relay inputs (currents, voltages or combinations) are purely sinusoidal and the relay response is checked. In dynamic testing the sinusoidal inputs are superimposed with dc offsets and harmonics. The response is again noted. Needed for testing are steady-state test benches, dynamic test benches and routine maintenance.

Relay testing at regular intervals relieves the concern of the operating personnel. Yet overtesting by different groups tends to cause more concern. Hence, optimal testing intervals are required so that there is neither anxiety nor overconfidence in the protection equipment. There is always a certain satisfaction that the backup will take care of reduced maintenance, but, of course, at the cost of extensive dislocation to the power system.

11.2. STATIC TESTING

11.2.1. Secondary Injection Tests

Secondary injection tests and the equipment necessary are generally described in the manufacturer's manual. Brief details follow.

For testing single-input relays, one can derive current or voltage from the laboratory power supply. The current source is essentially a voltage source with a large resistance in the circuit. Any variation in the impedance of the relay circuit does not significantly change the current magnitude.

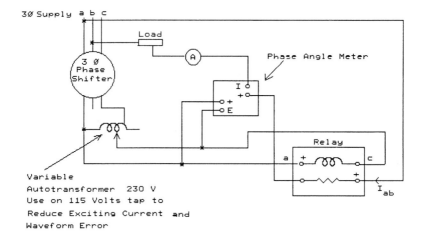

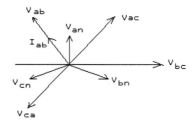

Figure 11.1 Single-phase static test circuits.

For testing two-input relays (directional, differential, distance, etc.), one requires

1. A variable current source with ammeter accurate enough on all ranges.
2. A variable voltage source with voltmeter accurate enough on all ranges.
3. A 360°, three-phase, phase shifter to change the phase shift between voltage and current with a resolution of at least 1.0°. Output voltages 110 V and $110/\sqrt{3}$ V.
4. Phase angle meter to within 1.0° accuracy.
5. A time interval meter with various starting and stopping modes accurate to within 1.0 msec.
6. Phantom loads in the form of high-wattage rheostats whose resistances remain substantially constant (the contact resistances of the brushes cause concern).

Chapter 11   291

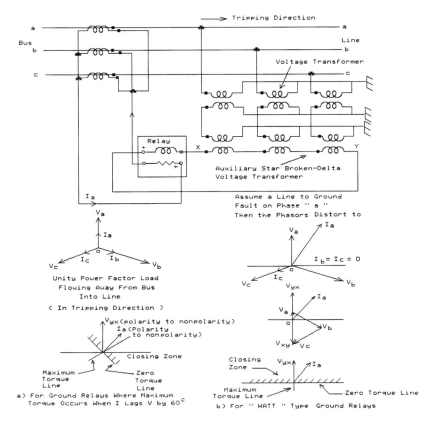

Figure 11.2 Test bench for ground-fault relays.

Figure 11.1 shows a single-phase test circuit for a relay to be energized from a voltage and current source. The input variations are the magnitudes of voltage and current and the phase shift between them. The voltmeter, ammeter and phase angle meter values are to be noted. Applying current and/or voltage to the relay may start a timer, and the relay trip outputs will stop the timer. The application of the relay inputs is invariably done by a high-speed contactor or molded case circuit breaker, and the auxiliary contacts are used to start the timer.

For tests on a ground-fault relay with zero-sequence polarization, the test circuit is shown in Fig. 11.2. Note the requirement for a star/broken delta transformer to generate zero-sequence voltage.

Figure 11.3 shows how to test a voltage-polarized ground-fault overcurrent relay, using a simulated ground-fault test. Note that the phase A

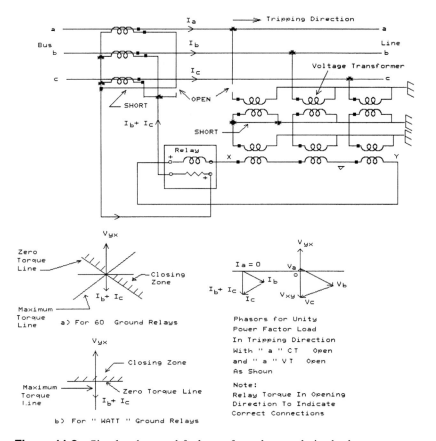

Figure 11.3 Simulated ground-fault test for voltage-polarized relay.

current transformer is short-circuited on the secondary side and opened on the relay coil side. Similarly, the phase A potential transformer is open-circuited on the primary side and short-circuited on the secondary. The ground-fault relay can be tested for all types of faults by shorting and opening the various CT and PT secondaries.

Theoretically, distance relays are required to measure the fault impedance, regardless of the magnitudes of voltage and current fed to it. In practice, the distance relay reach is affected by the magnitudes as well, as explained in Chapter 4. When a fault takes place on the line, voltages tend to drop and currents tend to increase. These changes essentially depend on the source impedance or the impedance behind the relay location.

Chapter 11

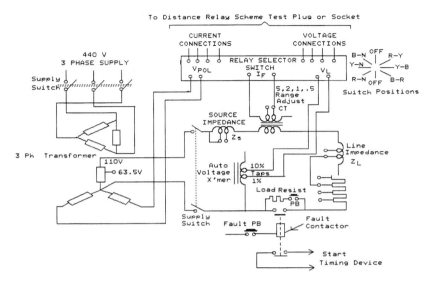

Figure 11.4 Static test bench for distance relays.

Figure 11.4 shows a circuit for testing distance relays that gives a more realistic simulation of the fault conditions. Basically, in the test circuit the source-to-line impedance ratio can be altered to see its effect on the variation of reach from the actual setting.

11.2.2. Primary Injection Tests

In a *primary injection test* the entire circuit (CT secondaries, relay coils, trip and alarm circuits, and all intervening wiring) is checked. Primarily injection is usually carried out by means of a portable injection transformer (Fig. 11.5), usually about 10 kVA with a ratio of 250 V/10 + 10 + 10 + 10 V.

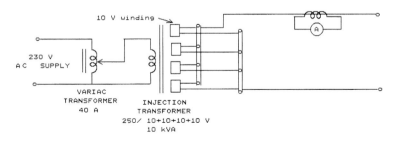

Figure 11.5 Primary current injection test.

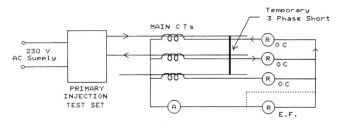

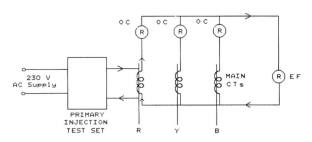

Figure 11.6 Test circuits for overcurrent relays.

This permits currents up to 1000 A to be obtained with all four 10-V secondary windings in parallel. The series combination of the four secondaries delivers up to 250 A. The injection current is normally controlled by a heavy-duty (about 40 A) autotransformer. Figure 11.6 shows the test circuits for both phase-fault and ground-fault overcurrent relays. The test circuit for phase-fault directional relays is in Fig. 11.7. The relay contacts should close when the load current is in the tripping direction; they should open when power flow is reversed.

The test circuit for a ground-fault directional relay is shown in Fig. 11.8. The ground-fault directional relay does not operate under normal load conditions. The simulation is therefore done as follows: the primary of one phase of the voltage transformer is disconnected and its secondary shorted on itself; the secondaries of two current transformers are then shorted on themselves and the outgoing lead wires are open-circuited. In this simulation the ground-fault directional relay should close its contacts on power flow in the tripping direction and open on reverse power flow.

Figure 11.9 shows the directional check on a MHO relay with normal load currents. Increase the setting of zone 1 and zone 2 to maximum by reducing the restraint (i.e., increasing the resistance of the restraint circuit)

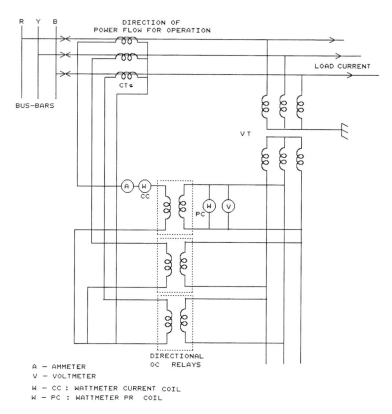

Figure 11.7 Test circuit for phase-fault directional relays.

and note the tripping for power flow in the tripping direction. The relay will open its trip contacts for power flow in the nontripping direction. Certain minor adjustments on the load current and its power factor are essential for this test.

The test circuit for a generator differential relay is shown in Fig. 11.10 and employs a primary injection test. Primary current is passed through one of the main current transformers and slowly increased until the relay operates. This gives the true effective value of minimum fault current necessary for relay operation and takes into account the CT magnetizing current and the relay pickup value.

The sensitivity of the transformer-biased differential relay can be checked with the primary injection set in a similar manner for generator protection. Figure 11.11 shows the test circuits for earth and phase-to-phase faults. Note that for simulation of a ground fault, the current is injected in

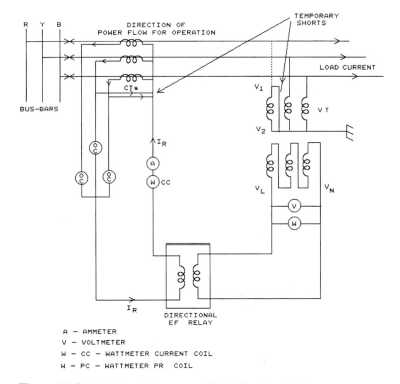

Figure 11.8 Test circuit for ground-fault directional relay.

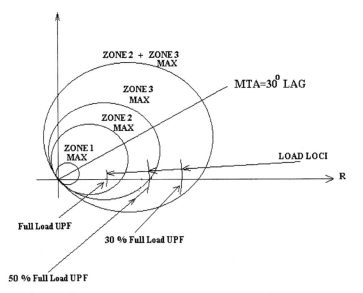

Figure 11.9 Directional check on an MHO relay.

Chapter 11

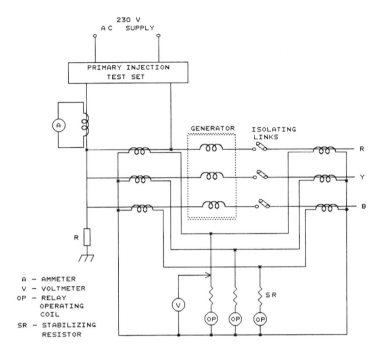

Figure 11.10 Test circuit for generator differential relay.

one current transformer only, whereas for simulating a phase-to-phase fault the current has to be injected in two current transformers.

Figure 11.12 shows the test circuit for a restricted earth-fault relay both for sensitivity, which is done by injecting the current through each of the main current transformers, in turn, and stability, which is checked by injecting the current through the neutral current transformer and each phase current transformer, in turn.

The test circuit for a sensitivity check of a high-impedance bus bar differential scheme is shown in Fig. 11.13. To check the correct value of the stabilizing resistance, a voltmeter must be used and V_{set} should agree with the calculations, within a reasonable limit.

The sensitivity test should be carried out for all three relays and the overall check relay. This should correlate with the minimum value of internal fault current for which the relay operates. This value essentially depends on the pickup setting rather than on the stabilizing resistance.

Figure 11.14 shows the test circuit for a negative phase sequence relay. Negative sequence relays are calibrated for negative sequence current only,

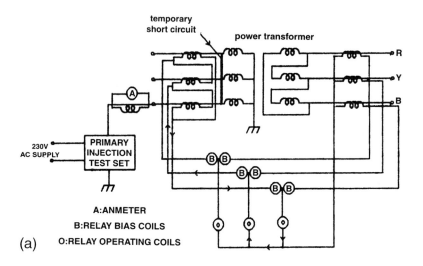

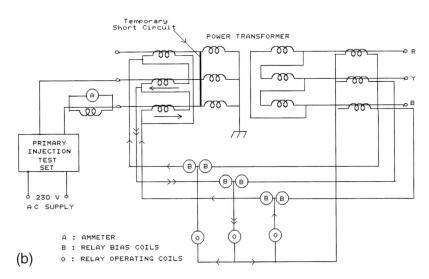

Figure 11.11 Test circuit for biased differential relay for transformers: (a) earth faults; (b) phase faults.

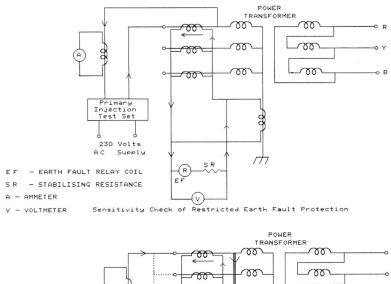
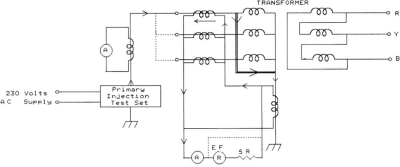

Figure 11.12 Sensitivity and stability of restricted earth-fault relay.

whereas the simulation in the test circuit is for a phase-to-phase fault. Therefore

$$1/\sqrt{3} \times \text{injected current} = \text{negative sequence current}$$

11.3. DYNAMIC TESTING

The primary and secondary injection tests are primarily steady-state tests. The input voltage and current waveforms are purely sinusoidal, whereas in practical situations they may consist of dc offset and harmonics. The relays, particularly static and processor based, respond in various ways. Cost permitting, dynamic tests are highly recommended, although not mandatory.

300   **Test Procedures**

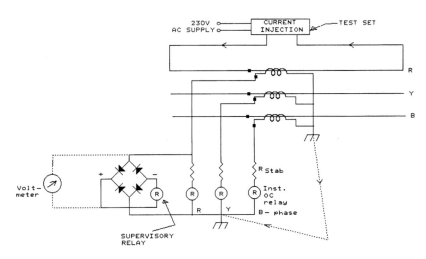

Figure 11.13   Test circuit for sensitivity check and stability of bus bar differential scheme.

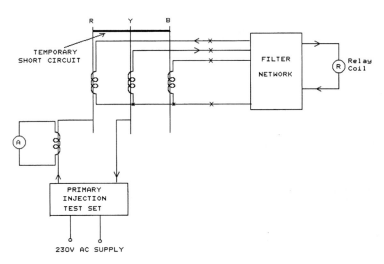

Figure 11.14   Test circuit for negative sequence relay.

Almost all relays can be tested against the dynamic test bench for accuracy, through-fault stability and operating time. The bench consists of a three-phase source, variable source impedance, variable line impedance, X/R ratio, fault switch and selection of instant of fault on the voltage waveform. Figure 11.15 shows the line diagram of the three-phase dynamic test bench.

The dynamic test bench is basically a three-phase circuit consisting of impedances representing the source generation levels, transmission line constants, load impedances, fault impedances and earth impedance. All these are independently variable, so the fault MVA linked with the system voltage level, fault time duration, system Z_0/Z_1 ratio, and source, line and system X/R ratios, etc., may be completely satisfied for the particular test envisaged.

The artificial three-phase line is supplied by a 440-V, three-phase, 60- or 50-Hz mains. The maximum fault current for which the bench is designated is 100 to 150 A (short time). The continuous rating is around 30 A (30 times 1.0 A). For testing 5.0-A relays the specifications of the bench need revision as follows:

1. Continuous rating: 150 A (30 times 5.0 A).
2. Short-time rating: 500 A or more.
3. Source impedance: 2.0 to 144 ohms in 24 equal steps; X/R ratio = 10 or 20 or 30.
4. Line impedance: 0.1 to 10 ohms in 24 equal steps; X/R ratio = 1.0, 1.73, 2.75 and 10.
5. PT ratio: 440/110 line to line.
6. CT ratio: Variable/secondary 5.0 and 1.0 A; primary turns, 50 with tap at every 5 turns; secondary turns, 150 with tap at 30 turns (1.0 and 5.0 A); auxiliary secondary, 33 turns with tap at every turn. Thus the CT ratio is variable in wide limits with secondary rated at either 1.0 A or 5.0 A. Knee-point voltage at 300 turns is 1000 V.
7. All faults (three-phase, L-L-G, L-L and L-G faults) can be made by a high-speed contactor, to be energized at any instance on the voltage waveform. This is called a *point-of-wave fault switch*. Thus, any amount of dc offset can be generated in the fault current, even in the relay voltage.
8. For creating arcing faults, a resistance of 4.0 ohms, in steps of 0.1 ohm, can be inserted in the fault.
9. The timing, such as energization of the bench, triggering the oscilloscope, fault duration, etc., including the measurement of relay operating time, is accomplished by a digital timer of 10 sec with a lowest time of 0.1 msec.

Test Procedures

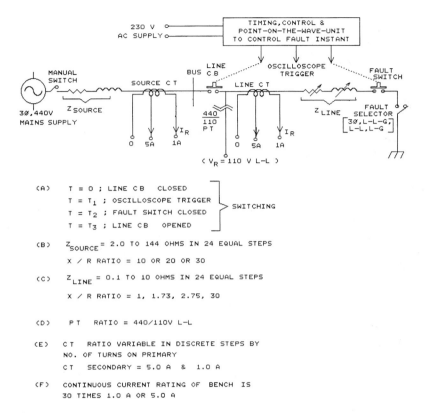

Figure 11.15 Dynamic test bench.

The following protection equipment can be tested under static and dynamic conditions on the dynamic test bench.

1. Distance relays
2. Generator differential protection
3. Transformer differential protection
4. Overcurrent relays
5. Phase comparison, etc.

11.4. COMPUTER-BASED TESTING

A digital computer has a tremendous amount of programming flexibility, and when its output is interfaced with a high-current, high-power amplifier, one can generate any programmable waveform of voltage and current signals, against which the relays can be tested for sensitivity and stability. The block diagram of one such test module, based on a personal computer, is shown in Fig. 11.16.

The test unit normally comprises

1. A universal three-phase rig, incorporating current and voltage amplifiers, measurement section and control electronics
2. An IBM-compatible PC with DOS operating system
3. Testing software for various relays

All device functions are controlled by the PC operating software. Instructions are entered with a mouse or keyboard, and data are printed out with an external printer.

Communication between the PC and test rig is established by using input/output ports. The relay under test is connected to the test unit via connectors or banana sockets. The menu-driven user friendly interface can be operated without prior programming knowledge.

The basic features of the PC-based test rig are as follows:

1. It provides three currents and three voltage outputs. All outputs are continuously and independently adjustable in amplitude, phase and frequency. No range tapping is necessary.
2. All current and voltage outputs are protected against overload and short circuits.
3. AC 60- or 50-Hz mains are galvanically isolated from the test unit.
4. A precision measurement section is available.

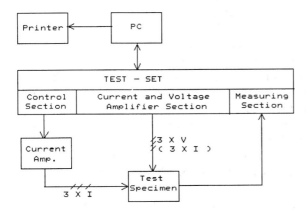

Figure 11.16 Computer-based test bench.

5. Two or more potential-free contacts for external switching are provided.
6. Simulation of all faults can be performed.

11.5. MAINTENANCE SCHEDULE

The performance figure of the protective equipment, with proper maintenance, is 95% or even more. The remaining uncleared faults or malfunctions are due to

1. Limitations of the protective relay itself
2. Faulty relays
3. Defects in pilot wires
4. Incorrect connections
5. Incorrect settings leading to loss of selectivity, etc.

Invariably, the failure of protective equipment to operate properly is due to its setting at the time of commissioning, which may have been 10 or 20 years earlier. Meanwhile, considerable changes could have occurred in the network, leading to changes in fault levels, equipment, parameters, etc. Thus, it is advisable to check the correctness of the settings at specified time intervals. For example, in a high-impedance bus bar differential scheme, the through-fault level might have increased as a result of added generation or lines. If the stabilizing resistance is not corrected, the relay will malfunction for an external fault.

It is desirable to inspect and test the protective gear, including circuit breakers (e.g., CB contacts are likely to be welded during reclosure on permanent faults), about every 6 to 12 months. This ensures a desirable confidence level among protection personnel. Too frequent testing should be avoided because it creates loss of confidence among protective engineers.

Sometimes the gear must be checked on line when network load conditions are low. The tripping contacts of the gear under test must be isolated and, for the test duration, remote backup must be used. The simulated test can now be carried out. For example, consider the sensitivity of a bus differential scheme for an internal fault. Sensitivity can be checked by shortening a set of CT secondaries and opening the pilot wires feeding the relay. Since the balance current from the shorted CT is not available, the differential relay is fed with the line current (whose CTs have been shorted). The correct operation of the relay can be checked. This test checks every item from CTs to relay.

Finally, off-line insulation tests using a 1000-V ac megger should be carried out on the relay pilot wires to check for insulation deterioration.

11.6. PROBLEMS AND EXERCISES

1. Draw the single-phase test circuit for experimentally plotting the polar characteristics of various distance relays.
2. Draw the circuit, and explain its operation, for deriving a high current from mains for testing overcurrent relays.
3. A bus has two incoming lines and one outgoing line. It is desirable to simulate an internal fault on the outgoing line to check the sensitivity and correct operation of the high-impedance bus differential scheme. Explain what changes have to be made on the CT secondary of the outgoing line.

12
Recent Advances and Futuristic View

12.1. INTRODUCTION

Most of the relays discussed have depended on fundamental frequency voltages and currents for correct operation. A fault increases current magnitude and decreases voltage magnitude and changes the phase shift between them. The fault also causes transients or noise, such as dc offset, harmonics, distortion in the electrostatic and magnetic fields surrounding the power system element, traveling waves, etc. Transients may lead to relay malfunction, where the relay is assumed to be fed only with sinusoidal quantities. Thus, any event or fault-generated noise needs to be removed or filtered from the relaying voltages or currents and then fed to the relay.

Some protection engineers therefore posed the following problem: Develop relays that operate on noise rather than on the fundamental frequency components at the relay location. These relays ought to be faster than those based on fundamental frequency relay inputs.

Recently developed relays use traveling wave or harmonics. Other developments include

Dynamic relays, which are augmented by the first, second or both derivatives of the relay inputs. Use of these relays is called *dynamically shaping* the relay characteristics. Such a relay has different transient and steady-state operating characteristics.

Adaptive relays, which adapt to variable fault characteristics, depending on the prefault power flow, say, for example, the fault characteristics of the line swivels depending on prefault export or import of power. The relay characteristics must swivel to fit around the variable line fault characteristics, thus ensuring minimum malfunction.

Harmonic-based relays. A ground-fault relay based on harmonics has been reported for synchronous machines.

Statistical relays, such as relays for overhead line protection against a high-impedance ground fault. A high-impedance ground fault does not lead to a large fundamental frequency fault current, and it virtually goes unnoticed. The high-impedance ground fault invariably is due to breakage of a phase conductor of medium voltage falling on dry ground. The resultant touch and step voltages lead to electrocution.

Intertripping without a carrier. For high-speed, single-shot autoreclosure, the CBs must trip simultaneously and quickly from both ends of the high-voltage line. It is normal practice to employ either carrier-aided distance schemes or unit carrier schemes, both of which require a carrier (50 to 500 kHz) for transferring information from one end of the line to the other.

12.2. RELAYS BASED ON TRAVELING WAVES

The work reported here (protection of ac and dc high-voltage lines) is under intense investigation. Modern EHV lines require faults to be cleared rapidly and selectively. The major problem encountered in reducing relay operating time is that, immediately after the fault, traveling waves distort the sinusoidal voltage and current at the relay location.

Most methods of determining the fault position on a transmission line are based on the fundamental frequency voltage and current waveforms. Consequently, the undesired transient signal components (i.e., noise) must be filtered out before the fault impedance is determined.

In solid-state static relays noise is removed by analog filtering, whereas in a number of computer-based real-time distance protection algorithms digital filtering is used. Both methods cause an annoying time delay, thereby defeating the main purpose of reducing response time.

This section describes a single-phase version of ultra high-speed directional relay, responding to the direction of traveling waves, created by a fault and therefore suitable for a directional unit carrier scheme. The traveling-wave directional relay derives its actuating quantity from a short stretch of conductor, called a *coupler*, mounted below the power line. The coupler thus has simultaneous electric and magnetic coupling with the power conductor. This coupling replaces both current and voltage transformers.

12.2.1. Traveling Waves for Internal and External Faults

A transmission line between bus A and bus B with assumed positive direction of wave propagation for the two relay locations is shown in Fig. 12.1.

Chapter 12

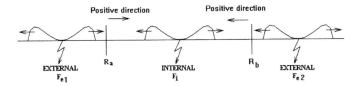

Figure 12.1 Traveling waves for internal and external faults.

As soon as a fault occurs on the transmission line, traveling waves of magnitude equal to the prefault voltage at the fault point propagate in both directions and pass over the relay locations. For the two external faults and one internal fault, the direction of propagation, as viewed from relay locations R_A and R_B, is summarized in the following table.

	Direction of traveling waves		
Fault	Relay R_A	Relay R_B	Trip/no trip
Internal F_i	Negative	Negative	Trip
External F_{e1}	Positive	Negative	No trip
External F_{e2}	Negative	Positive	No trip

In the directional unit carrier scheme, R_A and R_B are required to detect the direction of wave propagation and communicate with each other over a carrier channel to take a trip or no-trip decision. In the worst case of a fault near any bus, the scheme has an operating time of

T operating = time required for the traveling wave and the carrier to propagate over the complete line.

At the speed of light (300,000 km/sec), the relay for a 300-km line would operate in 2.0 msec.

12.2.2. Traveling-Wave Directional Relay

The directional relay takes the form of a coupler terminated at both ends by resistors R_1 and R_2, as shown in Fig. 12.2. The coupler has a mutual capacitance C_{12} and mutual inductance M_{12} with the main power line.

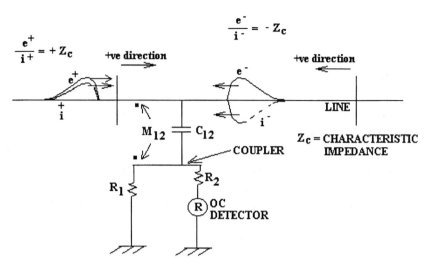

Figure 12.2 Coupler as directional relay.

The next section proves that, by appropriate choice of R_1 and R_2, the current in the overcurrent detector is zero for a forward traveling wave e^+ and finite for a reverse wave e^-. The overcurrent detector, therefore, serves the purpose of a directional detector, capable of detecting whether the fault is in the forward or reverse direction.

12.2.3. Operating Principle

The capacitive coupling between the power line and the coupler is shown in Fig. 12.3, the magnetic coupling in Fig. 12.4. The capacitive coupling

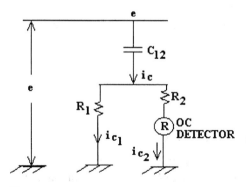

Figure 12.3 Capacitive coupling.

Chapter 12

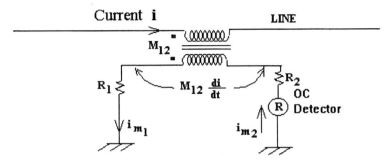

Figure 12.4 Inductive coupling.

current is flowing in the same direction through R_1 and R_2, whereas the inductive coupling current flows in opposite direction in R_1 and R_2.

The following analysis shows how the current in the OC detector can be made zero for a traveling wave in the nontripping direction and finite for a traveling wave in tripping direction. For capacitive coupling (Fig. 12.3) the currents are

$$i_c = C_{12} \frac{de}{dt}$$

$$i_{c1} = \frac{R_2 C_{12}}{R_1 + R_{12}} \frac{de}{dt}$$

$$i_{c2} = \frac{R_1 C_{12}}{R_1 + R_{12}} \frac{de}{dt}$$

For magnetic coupling (Fig. 12.4) the currents are

$$i_{1m} = i_{2m} = \frac{M_{12}}{R_1 + R_2} \frac{di}{dt}$$

The resultant currents are then:

Current in $R_1 = i_1 = i_{c1} + i_{m1}$

$$= \frac{R_2 C_{12} \frac{de}{dt} + M_{12} \frac{di}{dt}}{R_1 + R_2}$$

Current in $R_2 = i_2 = i_{c2} - i_{m2}$

$$= \frac{R_1 C_{12} \frac{de}{dt} - M_{12} \frac{di}{dt}}{R_1 + R_2}$$

For forward and reverse traveling waves on the transmission line, the following relation exists:

$$\frac{e^+}{i^+} = +Z_0 \quad \text{for forward traveling wave}$$

$$\frac{e^-}{i^-} = -Z_0 \quad \text{for reverse traveling wave}$$

For a forward traveling wave, where $e = iZ_0$,

$$i_1 = \frac{1}{R_1 + R_2}[R_2 C_{12} Z_0 + M_{12}]\frac{di}{dt}$$

$$i_2 = \frac{1}{R_1 + R_2}[R_1 C_{12} Z_0 - M_{12}]\frac{di}{dt}$$

For $i_2 = 0$,

$$R_1 = \frac{M_{12}}{C_{12} Z_0}$$

Thus, by proper choice of R_1, the overcurrent detector does not operate.

For a reverse traveling wave, where $e = -iZ_0$,

$$i_1 = \frac{1}{R_1 + R_2}[-R_1 C_{12} Z_0 - M_{12}]\frac{di}{dt}$$

The capacitive and inductive currents subtract in one OC detector, whereas they add in the other detector for a traveling wave in a particular direction. Thus, the OC detectors behave as directional detectors, capable of detecting whether the fault is in the forward or reverse direction. Such detectors at the ends of the EHV line, along with a carrier, constitute a unit carrier scheme.

Work is progressing on extending this work to three-phase lines. The expected solution involves stringing two couplers below the three-phase line, suitably terminated by resistances.

12.2.4. Conclusion

These sections have described a directional relay responding to the direction of propagation of traveling waves generated by a fault. The relay is, therefore, suitable for directional comparison unit carrier protection of EHV/UHV lines. Operating time is very small and equal to or less than the propagation time over twice the length of power line under protection.

Separate current and voltage transformers are not required, since their purpose is served by capacitive and magnetic coupling of the coupler. Extension of this work to three-phase lines is in progress.

12.3. RELAYS BASED ON STATISTICAL NATURE OF NOISE

12.3.1. Introduction

One of the most difficult problems, and not yet satisfactorily solved, is to detect a high-impedance ground fault on medium-voltage lines. Some examples are

A power conductor snapping and falling on dry ground
A power conductor contacting a tree branch
A polluted, partially conducting insulator

Existing protective relays (i.e., overcurrent) are not capable of detecting such faults, since the fault current is negligible. Although such a fault does not damage the line, humans and animals are jeopardized by the touch and step voltages, which can be fatal. This has motivated many protection engineers to think of an entirely new type of protection.

12.3.2. Principle of the Relay

The dominant property, identified on high-impedance ground faults, is arcing. To investigate this property, University of Texas A & M scientists staged a total of 86 separate faults on six different voltage lines. The most important property observed was increased high-frequency activity (Fig. 12.5). The frequency range was around 2.0 to 10 kHz, where each frequency was a certain percentage of the fundamental (i.e., 50 or 60 Hz). Switching loads also caused similar high-frequency activity, but the duration of time for the arcing fault was much longer than that due to switching.

The basic protection law was therefore postulated as: If the increase in the high-frequency activity (i.e., 2.0 to 10 kHz) lasts much longer than the duration of such activity in the event of switching, a trip decision is to be taken. The fault is then assumed to be a high-impedance ground fault.

The relay was built around a microprocessor, and its performance was satisfactory. Unfortunately, it did not have adequate reliability because adjacent lines also picked up this noise and, occasionally the relays of the healthy lines also operated. The matter is still under intensive investigation.

12.4. RELAYS AUGMENTED BY DERIVATIVES (DYNAMIC RELAYS)

Most of the relays discussed operated on steady-state, sinusoidal input voltages or currents or both. The relay operating characteristics were steady-

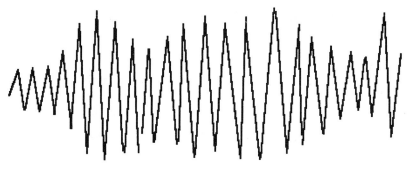

UNFILTERED FAULTED PHASE CURRENT

2-10 kHz FAULTED PHASE CURRENT

Figure 12.5 High-frequency activity in relay inputs consequent to high-impedance ground fault.

state characteristics. They did not take any account of the rate of change of the input quantities (i.e., time derivatives of current, voltage, impedance, etc.).

It is known that relay inputs change suddenly during a fault, whereas under other conditions (e.g., power swing, changing load) they change gradually (i.e., in time). It was felt that if the trip criterion or the control law for the protective relay were augmented by the derivatives of the inputs, the relay characteristics would be dynamic or dynamically changing when time derivatives exist, and conventional static or steady state when the time derivatives are zero.

Take, for example, an MHO distance relay (Fig. 12.6). The figure shows the MHO characteristics on the R-X diagram with maximum torque angle of 90°. The characteristics plot as a circle with radius $X_n/2$ and center coordinates $R = 0$ and $X = X_n/2$. The trip equation or the inequality relationship for the MHO relay is

Chapter 12

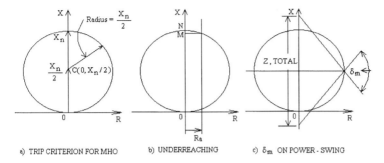

Figure 12.6 The MHO distance relay.

$$R^2 + \left(X - \frac{X_n}{2}\right)^2 \leq \left(\frac{X_n}{2}\right)^2$$

Inside the circle is the trip region, the restraining region is outside. The figure also shows the behavior of the relay on arcing faults and power swings. If R_a is the fault resistance the relay underreaches by an amount MN. The maximum power angle which the relay can accommodate without malfunctioning is also shown. Now in the event of a fault, the prefault load impedance Z_L instantaneously falls to Z_F. In the event of a power swing, the impedance moves along the locus of the power swing rather gradually compared to the fault.

Our purpose is to enlarge the MHO characteristics along the R-axis so that the relay can accommodate a large fault resistance, particularly for high-impedance ground faults, and to constrict the relay characteristics, under power swing conditions, so that it accommodates a large power swing angle δ. This ensures that the relay does not malfunction during a stable power swing from which the system is capable of recovering (i.e., no loss of synchronism) by itself. This feature eliminates the need for an additional out-of-step blocking relay. Only an out-of-step tripping relay is needed.

12.4.1. Choice of Derivatives

Figure 12.7 shows the power swing locus and load point P along with MHO characteristics. The power swing is initiated form point P, penetrates a maximum P_m the MHO characteristics and returns, if the power swing is stable. A conventional MHO relay tends to malfunction on a stable power swing unless additional out-of-step blocking is incorporated.

The variation of resistance R, its first time derivative $\dot{R}$ and the second derivative $\ddot{R}$, with respect to time, are shown in the bottom portion of Fig.

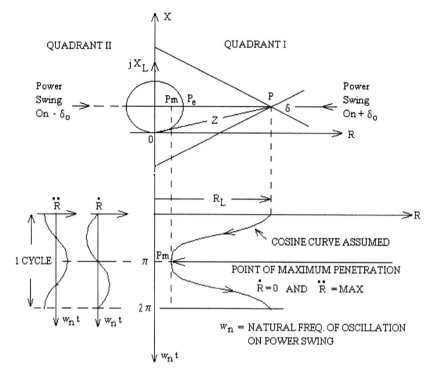

Figure 12.7 R, $\dot{R}$ and $\ddot{R}$ on power swing.

12.7. At maximum penetration of power swing, $\dot{R}$ tends to be zero, whereas $\ddot{R}$ tends to be maximum. Our aim is to have maximum constriction of the relay characteristics at this point. Therefore, it is desirable to augment the relay trip criteria, or control law, by $\ddot{R}$ of the apparent resistance $Re(V/I)$ seen by the relay.

12.4.2. Augmented MHO Relay

The revised trip criteria for the MHO relay, augmented by $\ddot{R}$, is now

$$(R + W\ddot{R})^2 + \left(X - \frac{X_n}{2}\right)^2 \leq \left(\frac{X_n}{2}\right)^2 + (W\ddot{R})^2$$

and

$$(R - W\ddot{R})^2 + \left(X - \frac{X_n}{2}\right)^2 \leq \left(\frac{X_n}{2}\right)^2 + (W\ddot{R})^2$$

See Fig. 12.8. For $\ddot{R} \geq 0$, the first equation generates a circle with X_n as a chord. The center of the circle is shifted left along the R-axis by an amount $W\ddot{R}$ and the radius increases by $(W\ddot{R})^2$, so X_n becomes a chord. The second equation generates a similar circle displaced to the right.

Due to logical and compounding of the two equations, the resultant characteristics are the common area of the circles with X_n as the common chord. Thus, on a power swing, due to augmenting by $W\ddot{R}$ the MHO circle constricts itself, thereby making itself considerably immune to possible malfunction on power swing.

For $\ddot{R} \leq 0$, the two circles generated by the two trip equations interchange and the resultant characteristics remain unaltered. The constriction of the MHO characteristic is, therefore, irrespective of the polarity of $\ddot{R}$. In effect, it means that the constriction of the augmented MHO relay is independent of whether the power swing moves from quadrant I to quadrant II, or vice-versa.

12.4.3. Block Diagram

Figure 12.9 shows the block diagram of the MHO relay, where X, R and $\ddot{R}$ are shown as inputs computed by any of the processor or computer-based distance relay algorithms. The decision block, which compares $\ddot{R}$ with the set value, decides whether there is a fault or a power swing. In a fault the trip equations of the previous section are not augmented by the derivatives (weighting factor $W = 0$). The relay therefore will clear high-resistance faults.

In a power swing the trip equations are now augmented by $\ddot{R}$ and the relay characteristics constrict themselves so as to be immune to the power swing.

12.4.3.1. The Decision Block

The set value of $\ddot{R}$ in the decision block, which decides the switching in or switching out of augmentation, depends on its value under a fault and a power swing. Under fault conditions the prefault resistance component of impedance seen by the relay virtually falls to zero instantaneously, resulting in almost impulse or a very large value of $\ddot{R}$.

Under power swing conditions, $\ddot{R}$ needs to be evaluated for a given system, although rough estimates can be made if the assumption that the predisturbance resistance value of $R = R_L$ in Fig. 12.7 falls to virtually zero (i.e., power swing enters almost to electrical center of the network) in a cosine fashion in a half-cycle period, corresponding to natural frequency of oscillations given by the electromechanical swing equation.

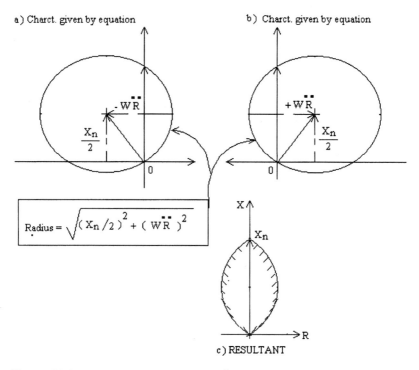

Figure 12.8 MHO relay augmented by $\ddot{R}$.

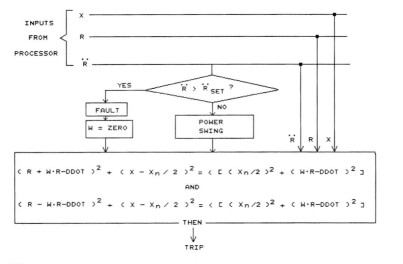

Figure 12.9 Block diagram of the augmented MHO relay.

In Fig. 12.7 let

$$R = R_L \cos w_n t$$

where $w_n = 2\pi f_0$. The value of f_0 may be approximately taken as 1 to 3 Hz. Differentiating twice we get

$$\ddot{R}_{max} = -R_L w_n^2$$

The set value, therefore, in the decision block must equal more than this value.

12.4.3.2. Weighting Factor

The larger the weighting factor W, the more the relay characteristics are constricted under power swing conditions. The characteristics (i.e., and compounding of two circles) vanish if

$$W\ddot{R}_{max} \geq \frac{X_n}{2} \quad \text{or} \quad W_{max} \geq \frac{X_n/2}{\ddot{R}_{max}}$$

where $\ddot{R}_{max} = -R_L w_n^2$.

12.5. SWIVELING DISTANCE RELAYS (ADAPTIVE RELAYS)

As discussed in Chapter 4:

> The quadrilateral fault characteristics of the line tend to swing clockwise on prefault export conditions.
> They tend to swing counterclockwise for prefault import conditions.

It is therefore desirable to automatically swivel the relay characteristics as well. To achieve this, zero-sequence current compensation of the relay inputs has been tried, but the results have not been satisfactory and the problem needs further investigation.

12.6. SOFTWARE FOR RELAY SETTING

Several organizations have developed powerful interactive software packages for load flow, fault analysis, stability analysis and relay coordination package (RECAP). The RECAP package contains program control, data and result files, a device library of relay data, a display of fault results on an R-X diagram for checking both distance relay and OC relay settings.

12.7. INTERTRIPPING WITHOUT A CARRIER

In conventional distance protection only the middle 60% of the line gets high-speed protection. This is inadequate for high-speed single-shot autoreclosure in EHV lines. High-speed protection to 100% of the line has traditionally been done with the help of a carrier. This is discussed in detail in Chapter 5.

A recent approach to this problem proposes eliminating the carrier. This is possible by making better use of information available at the local end. The concept can be described as follows. Consider the transmission line LM with a source at each end and protected by an impedance relay Z installed at end L (Fig. 12.10). The time sequence of events is

> Normal load flow.
>
> Fault occurs at point P in zone 2 as seen from end L. Currents and voltages at the relay location change from their healthy values. This is termed the first change.
>
> Fault is cleared from end M, in zone-1 time (instantaneously). Hence, the voltages and currents at the relay at L change once again. This is called the second change.

The second change in negative sequence and zero-sequence current components at the relay location allows us to draw inferences about remote CB operation and to trip the CB at L (Fig. 12.10). The decision is arrived

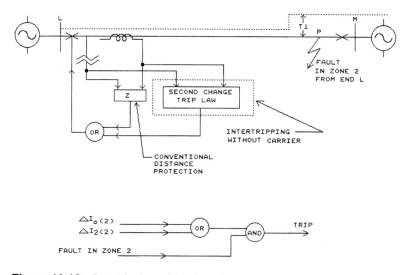

Figure 12.10 Intertripping without carrier.

at with the help of information available at end L only, thereby doing away with the carrier altogether. It can be thought of as an intertripping without carrier. The scheme, however, is fraught with certain problems, among them:

Low sensitivity for systems where second change is small.

Inability to distinguish between some developing faults in the adjoining zone from faults within the protected zone. (Developing faults are those which initially start as L-G faults but eventually end up as L-L-G faults.)

Malfunction when the next line is a short line with pilot protection.

Attempts are being made to overcome the above lacunae by suitably compensating the second-change current quantities with additional healthy phase voltages. The scheme appears to be particularly attractive in view of the tremendous processing power that the present relay has at its command. There in no reason why even more complicated discrimination cannot be formulated and implemented for intertripping without a carrier.

12.8. FUTURISTIC VIEW

12.8.1. Artificial Neural Network

Refer to Figs. 12.11 and 12.12. What is an artificial neural network (ANN)? An ANN is a computer architecture inspired by research about the human brain. It attempts to build computing systems analogous to the biological nervous system. Its basic constituents are

1. The processing element, called *neuron*
2. Network topology
3. Learning (or training) method
4. Recall method

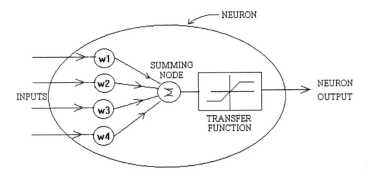

Figure 12.11 The neuron model.

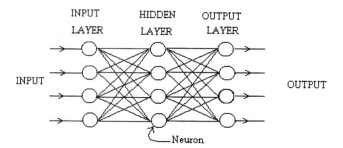

Figure 12.12 A typical multilayer neural network.

1. *The neuron*: This is the basic ANN building block. It consists of many input paths, called connections, a transfer function and one output. A neuron calculates its output by taking the weighted sum of all its inputs and passing it through the transfer function. Each neuron is completely self-sufficient and works in total disregard to processing within its neighbors. At the same time, all processing elements intimately affect the entire network behavior since each neuron's output becomes an input to many others.

2. *Topology*: ANNs are created by interconnecting neurons. The topology depends on application, network size and training method. Often an ANN is subdivided into layers: the first is the input layer, the intermediate layers are called hidden layers, and the final output is obtained from the output layer.

3. *Method of training* (learning): For the ANN to perform useful work, we have to train it with the help of an established set of input/output values. The learning law allows the processing element's response to change with time, depending on the nature of the input signal. It is the means by which the network adapts itself to the desired output and thus organizes the information within itself.

The training process progressively adjusts the weights between the neurons to enable the desired output to be achieved. The most widely used learning law is the *back-propagation law*. In this method the present output of neurons in the outermost layer is compared with the desired value, and the error is determined and allocated to the weights of all connections, from the output to the input layer recursively till the process converges.

4. *Method of recall*: This is the process of finding the output response for a given input to the network. Once the neural network has learned, the learning law may be disabled and the weights frozen.

One need not actually build a network in order to test its performance. Several software simulation packages are available which let the user simulate and train the network and experiment with various topologies and learning before committing it to hardware.

There are some problems, which basically fall under the category of pattern classification. These problems are peculiar in that it is difficult, or sometimes impossible, to express the relationship between the input and output mathematically, but an endless number of input/output patterns can easily be identified by humans. Neural networks attempt to capture such complex correlations with the help of a machine without explicitly programming it. This is not a von Neumann type of architecture.

Complex pattern recognition problems occur when an incipient or broken conductor fault, among others, occurs. Here an ANN may be successfully applied. ANNs, however, are not intended to replace conventional methods. At most they help us grapple with some very tricky pattern recognition problems.

Most reported work is at the simulation level. However, some vendors have come out with integrated circuits which will help build true hardware prototypes in the near future. The following work in the protection field, based on neural networks, has been documented.

> Detection of incipient faults on power distribution feeders
> Adaptive single-pole autoreclosure techniques for EHV lines
> Discrimination between magnetizing inrush and internal faults

12.8.2. Electric and Magnetic Field Distortion

Every power system element under healthy conditions has electric and magnetic fields surrounding it. An unbalanced fault distorts these fields, whereas a balanced fault collapses the electric field and builds up the magnetic field. There is no reason why relays cannot be built to detect changes in the fields. A traveling-wave relay, as discussed in Section II falls in this category.

12.8.3. Computationally Fast Orthogonal Functions

The advent of microprocessors ushered in a new phase in protection systems. These systems must employ computationally fast real-time algorithms for extraction of fundamental frequency components drowned in postfault noise. For this purpose the reported orthogonal functions are either Fourier or Walsh and a number of other functions. The algorithms based on these orthogonal functions slow down computationally, depending on the number of multiplications required. The multiplication instruction is computationally quite slow.

A new set of orthogonal functions called [1-shift] has been reported for extracting the fundamental component. So far it is the fastest orthogonal function with good frequency response. In sine- and cosine-type orthogonal functions all multiplications have been replaced by right and left shifts. If the number of right shifts is n, then the sample is divided by $1/2^n$. The shifting of numbers in the accumulator is much faster than in ordinary multiplication. The topic needs further investigation.

12.8.4. Aperiodic Sampling

In most of the known digital methods for extracting fundamental frequency components, the signals are sampled periodically and then correlated with sine- and cosine-type orthogonal functions. The correlation is basically done for integration purposes (discrete Fourier or Walsh integral, etc.). It occurred to Gauss that aperiodic sampling might have a more accurate result of integration. Hence, he formulated the following problem. If $\int_a^b f(x)\,dx$ is to be computed from samples of $f(x)$, where should these samples be taken to evaluate the integral to the greatest accuracy? In other words, how shall the integrating interval (a, b) be subdivided so as to give the best possible result?

The value of the integral defined by Gauss and Tchebycheff is

$$I = \frac{1}{n}[f(x_1) + f(x_2) + \cdots + f(x_n)]$$

where n = number of samples per cycle
$f(x_n)$ = value of nth signal sample

No multiplications are involved, so the integral is computationally fast and accurate, and thus gives excellent noise rejection. It has also been found that the points of sampling are not periodic but are symmetrically placed with respect to the midpoint of the integrating interval.

For example, for a sampling rate of six samples per cycle and integration limits of -1 to $+1$, the sampling instants are

$X_{-3} = -[0.433123] = -X_{+3}$

$X_{-2} = -[0.211259] = -X_{+2}$

$X_{-1} = -[0.133318] = -X_{+1}$

$X_{+1} = 0.133318$

$X_{+2} = 0.211259$

$X_{+3} = 0.433123$

These instants are shown in Fig. 12.13.

Note that the sampling is not periodic and can be easily implemented by a processor. This method of integration appears to be particularly suitable for real-time implementation.

Chapter 12

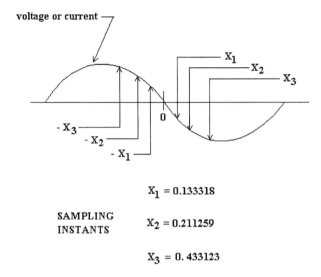

Figure 12.13 Aperiodic sampling instants for computation of integral.

12.8.5. Benchmark (Voltage and Current)

Year after year numerous new techniques of relaying are being documented. Some are solid state, and some are digital computer algorithms. Most of the authors rightly claim certain advantages for their techniques over older ones. Unfortunately, testing all new relay techniques and comparing them with known techniques does not appear satisfactory to this author simply because benchmark voltage and current waveforms, against which all techniques ought to be tested, do not exist. The time, therefore, has come to define benchmark voltage/current waveforms, at least for uniform testing purposes, for various power system relays.

12.9. LIMITS TO RELAYING

It appears that protection engineers have been spending their time and energy trying to improve relaying speed. For autoreclosure, a decrease in relay time means reduced CB interruption time. Parallel research work needs to be done on high-speed circuit breakers.

It appears that a half-cycle relay and a half-cycle breaker is, perhaps, the optimal solution. We have no control over reduction of deionizing time for arcing faults. The deionizing time perhaps shall dominate the speed of single-shot high-speed autoreclosure in high-voltage lines.

The relaying field is still as exciting and demanding—perhaps even more so—a century ago. New frontiers are being opened to relaying engineers in the protection of HV/EHV/UHV and HVDC lines. These will demand novel and creative approaches.

13
Central Computer Control and Protection

13.1. INTRODUCTION

Digital computers, initially used in electrical power systems for off-line power flow studies, short-circuit studies, stability studies, planning, load forecasting, etc., soon found use in real-time high-speed relaying, substation control, parameter monitoring, etc. On-line applications, involving time-critical and numerically intensive operations proved their reliability in a power system environment.

The advent of reliable cheap hardware, in terms of microprocessors and digital signal processing integrated circuits, enabled simpler tasks to be performed at distribution levels. Subsequently its use was widespread in protection and control functions. Thus, memory and high-speed mathematical and logical decision capacity are now available at the downmost level of "feeders." The easy communication between the other processor/computers provides access to data on other command levels and areas of control. These functions have now been transferred from hardware to very sophisticated software, thus making possible system adaptation and expansion by software configuration. As an example, addition of generation, transformers, transmission lines, etc., changes fault levels, thus requiring upgrading of the circuit breaker rating, new settings of protection relays, and so on. This is now being done automatically by software without much human intervention.

A major advantage of digital techniques is their ability to continuously self-monitor important circuits and functions to ensure the uninterrupted availability of the devices and, thus, quality of supply. All these innovations basically change the procedures of erection, commissioning, maintenance and system expansion. Reduction of maintenance, control and hardware monitoring has led to drastic changes in organizational structure and responsibilities.

328   **Central Computer Control and Protection**

Electric utilities need to take a fresh look into integrating various functions such as control, measurement, protection, metering and plant diagnosis. The many uses of voltage, current and power factor, the ability to correlate data from different sources or locations, due to improved communication, and many other possibilities support this new development. On the other hand, the utility has to ensure that vital and time-critical functions such as protection are not adversely affected by links with functions other than protection, in order to maintain protection accuracy, reliability, selectivity and operating time.

13.1.1. Hierarchical Structure

Because of the large distances (hundreds of miles) between generating plants and bulk consumer loads, the unpredictable demand of load by bulk power consumers, the safety requirements and the management of large integrated power systems needs to be organized on hierarchical levels. Typical control levels of such a structure, as conceived by Ungrad and many other researchers, are

1. Load dispatcher or central power system control
2. Regional load dispatcher or local power system control
3. Station control
4. Feeder or object control

A typical suggested control structure is shown in Fig. 13.1a. Asea Brown Bovri (Germany) calls the same strategy a *pyramid platform* concept (Fig.

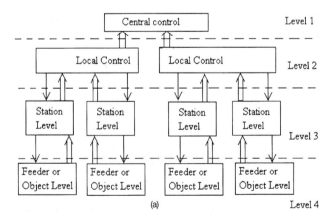

Figure 13.1 (a) Hierarchical structure for power system control ⇒ = data flow; → = command flow; (b) pyramid platform concept by Asea Brown Bovri; (c) Siemens telecontrol system.

Chapter 13

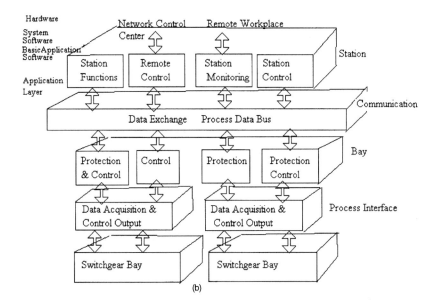

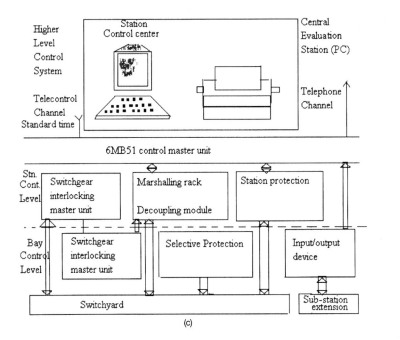

Figure 13.1 Continued

13.1b). The pyramid is a homogeneous concept for protection, monitoring, control and communication of all transmission and distribution substations. It claims to offer solutions from a single function protection relay or remote terminal unit (RTU) to fully integrated, complete, high-performance substation automation systems. Flexibility is provided to meet various user requirements. Adherence to open and published communication protocols enables the connection of other manufacturers' products and systems. Siemens (Germany) calls this strategy *telecontrol control system* (Fig. 13.1c). Regardless of the system, the typical tasks performed at central and/or local power system control levels are as follows.

13.1.2. Responsibilities at Different Control Levels

The tasks typically performed at the various hierarchical levels are summarized in Table 13.1. Data on communication channels flow mainly from the process via the feeder and station control levels to the local control level and commands in the opposite direction. In most cases the lowest level has the greatest priority.

Tasks fall approximately into three time periods (normal operation, fault, power system restoration) and differ considerably within these periods. Consequently, the requirements to be fulfilled by the communication and processing devices are basically different.

13.1.2.1. Normal Operation

The main task during normal operation consists of minimizing generation (thermal, hydro and nuclear) and transmission costs with the constraint of supply quality (i.e., voltage and frequency). Continuous supervision of various devices also forms part of the job.

13.1.2.2. Fault

The task is to perform protection functions and selective isolation of faulty power system elements from the healthy system as quickly as possible. This ensures minimum damage to power equipment due to exorbitant fault MVA and maintenance of stability in the grid, which thus ensures the highest possible reliability of supply to the consumers.

13.1.2.3. Power System Restoration

At the moment this task consists of islanding, intelligent load shedding, automatic CB reclosures for maintaining stability in the grid or burning out

Table 13.1 Tasks Performed at Central/Local Control Level

1	Normal 2	Fault During 3	Fault After 4	Restoration 5	Level 6
Exhange of data between computer systems on different power system control levels	●	●	●	●	3, 4
Data acquisition, verification and concentration	●	●	●	●	3, 4
Feeder protection		●			4
Bus bar protection		●			3
Backup protection		●			3, 4
Breaker backup protection		●			3
High-level supervision of protection settings		●		●	3
Autoreclosure			●	●	3, 4
Automatic switching of supply sources			●	●	3, 4
Generation control and load dispatching	●		●	●	3
Overload supervision of station, plant and lines	●		●	●	3, 4
Stability optimization	●		●	●	3
Automatic power system build-up switching procedures				●	3
Determination of switching sequences and interlocks	●			●	3
Event recording	●	●	●	●	3
Detection and recording of disturbances		●	●		3, 4
Fault location on overhead lines		●			4
Measurement and recording of analog signals	●			●	3, 4
Voltage regulation	●			●	3

Source: Ungrad et al., *Protection Techniques in Electrical Systems*. (Courtesy Marcel Dekker.)

semitransient faults on medium-voltage lines and, perhaps, many others. With extensive computerization, many other tasks can be thought of in the foreseeable future.

Table 13.2 lists typical control, protection and supervision tasks at the feeder and station control level.

Table 13.2 Control, Protection, and Supervision Tasks at Station Control Level

	State		
	Normal	Fault	Restoration
Exchange of data between computer systems on different power system control levels	●	●	●
Data acquisition and verification	●	●	●
Data concentration	●	●	●
Determination of system topology	●	●	●
Determination of load flow and voltages	●	●	●
Selection of power plants and generator units	●		●
Spinning reserves control	●	●	●
High-level frequency/active/reactive power regulation	●		
Operational reliability estimation	●		●
Estimation of the effects of power system changes	●		●
Event recording	●	●	●
High-level backup protection		●	
Systemwide confinement of faults		●	
Analysis of further development	●		●
Measurement and recording of selected parameters	●	●	●
Fault analysis			●
Reporting	●		●
Operator interfaces	●	●	●
Self-supervision and diagnosis	●		

13.1.3. Computer Structure in Substations

From Table 13.2 note that most important functions which have to be carried out quickly at the feeder level only need local information (e.g., current, voltage, protection configuration, measurement, metering). This especially applies to the fault state, which places high demand on processing time.

The feeder control level obtains, among other things, the setting for the protection devices and details of the system configuration from the station control level. The logical conclusion is, therefore, to construct switchgear bay units which integrate, as far as possible, all the functions associated with a bay in one or more microprocessors. Such special functions are protection, feeder control and interlocks, measurement, metering and diagnostics.

Whether it is possible to implement all these functions in a single processor or several parallel processors depends on the feeder voltage, operating philosophy, primary and backup protection, etc. The boundary conditions, with regard to the absolute immunity of the protection interaction and its priority in relation to other functions, are also highly important.

It also happens (frequently during transition from analog to digital technology) that some switchgear bay functions are performed by microprocessors and others, which belong to the same feeder, by conventional analog devices. This necessitates a suitable interface between the processor and analog devices. High-speed parallel processing is also the preferred practice in multiprocessor systems. See Fig. 13.2.

To what extent the bay units are equipped with their own MMCs (man-machine communication), thus permitting all settings, etc., to be performed by the feeder level, or all setting and supervisory actions have to be executed on a central MMC at the station control level, is a question of cost and operating philosophy.

The tasks at the bus bar and station control level are basically the same, namely bus bar protection and station control level process protection (bus bar and backup) and interlocking (i.e., bus-tie breaker, isolator) functions. The preceding remarks regarding multiple utilization of input variables and MMC also apply to units that serve as backup for the bay units.

The communication system between the bay and station control levels is arranged radially or as a bus. The highest reliability and speed demands it has to fulfill are determined once again by the bus bar protection, but synchronization and interlocking functions are also critical.

The basic requirements which digital protection and control systems have to fulfill with regard to selectivity, availability, speed, etc., are naturally

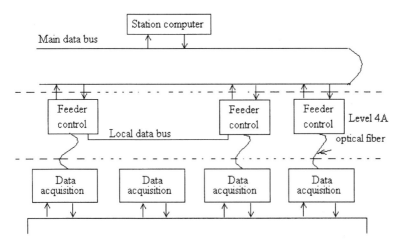

Figure 13.2 Typical computer structure for a substation.

the same as for conventional analog protection and control devices. Digital techniques, however, offer an additional advantage of extensive self-supervision and therefore open up new avenues for improving maintenance and availability.

Digital processing and measurement systems will come into their own as soon as the input variables are transferred in digital form by measurement transducers, such as conventional CTs and PTs. Strategies may change with the distinct possibility of newer types of current and potential transformers.

Fully integrated computer structures for substations are gradually being introduced. During the transitory phase from analog to digital relays and conventional secondary technology to a fully digital system, microprocessors will become standard practice in the individual protection, control and metering units, which will be applied together with analog devices.

Independent of its technical design, the computer-based control system receives its input signals from measurement transducers and transmitters and generates corresponding output signals for controlling the plant. See Fig. 13.3. The input signals may be analog, digital or binary states. Analog variables come mainly from CTs and PTs. Binary states originate primarily from the status (open or closed) of circuit breakers, isolators, tap changers, etc. Digital signals are transmitted by digital communication links.

The output signals generated by the computer-based protection and control system energize the actuating coils of circuit breakers, switches and power transformer tap changers. They change the settings of analog devices, are displayed on color monitor screens, logged on recording devices and

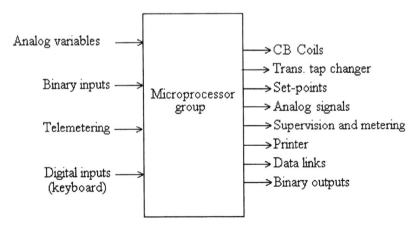

Figure 13.3 Control system for a substation.

Chapter 13

transmitted to remote locations by communication links. In some cases analog output signals provide continuous control of the plant.

13.1.4. Special Features of Digital Measurement and Control Methods

Converting analog variables to digital is quite different from processing an analog variable in an analog device. The major differences are the following:

1. Computer-based systems have an easily accessible, programmable and relatively large memory capability, but analog devices do not.
2. In fast control systems the objective is to execute a complete series of operations between individual samples, which is then repeated cyclically. This means that the computations have to be carried out in a very short time.
3. Computer-based systems are able to make better use of analog and binary state signals by determining their interrelationships.
4. Digital signals give no direct information of the behavior of the input variable between samples, which, compared with analog measurement, necessitates some complicated operations to determine zero crossings.

Therefore, there is little to be gained by simply mimicking analog protection and/or control devices digitally.

13.2. A/D CONVERSION OF ANALOG INPUTS

13.2.1. Introduction

A typical procedure for analog-to-digital conversion of input variables is given in Fig. 13.4. Continuously varying signals (usually sinusoidal currents and voltages from CT and VTs) are applied to the converter input. The signals, however, may not consist of just a 50- or 60-Hz fundamental, but may include superimposed high-frequency interference, harmonics, subhar-

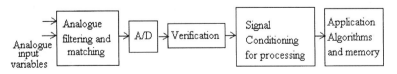

Figure 13.4 Procedure for converting analog signals.

monics and a dc component. The general equation for the input signal is thus

$$v(t) = V_0 e^{-t/T} + V_1 \cos(w_0 t - a)$$
$$+ \sum V_k \cos(kw_0 t - a_k)$$

where V_0 = initial value of dc component which decays at a time constant T
  V_1 = amplitude of the fundamental at rated frequency
  V_k = amplitude of kth harmonic

In the equation it is only the fundamental of rated frequency w_0 which is usually the most important. Before performing A/D conversion, it is important to ascertain which component in the input variable carries the desired information, and therefore should be converted accurately, and which is to be classified as interference, and thus suppressed.

13.2.2. Analog Signal Filter

In Fig. 13.5 the fundamental frequency component of $v = V_p \sin wt$ has a constant dc offset of V_0. If the sampling frequency equals the fundamental frequency, then

Sample value = $V_0 + v$

and thus has a constant error of V_0. If the sampling rate is twice the frequency of the ac component v, then

$$S_1 = V_p \sin wt + V_0$$
$$s_2 = V_p \sin(wt - 90°) - V_0$$

Thus, errors due to the dc component could be canceled by averaging the samples. Shannon's theorem clearly states that the sampling frequency should be at least twice the maximum frequency component expected in the input signal. To avoid errors while digitally extracting the fundamental, normal practice is the following:

1. The sampling rate should be at least four times the frequency of those components in the input variable which have to be accurately reproduced (i.e., $F_{sampling}$ = 4 × fundamental frequency).
2. The characteristic of the analog filter, preceding sampling and A/D conversion, should be such that it should filter out all higher frequencies more than $0.5 f_{sampling}$. This type of filter is called an *antialiasing* filter.

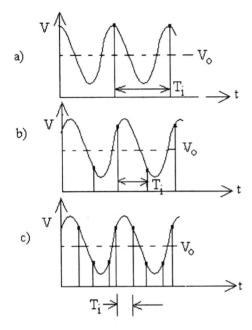

Figure 13.5 Influence of the sampling rate on the discriminating error.

The analog input variables applied to the A/D converter must be adjusted to lie within its operating range. A commonly used A/D converter accepts a bipolar input within range ±10 V peak.

13.2.3. A/D Conversion of Input Variables

Usually a single A/D converter chip with an analog multiplexer is used for several analog input variables. The purpose of the multiplexer is to apply the input variable to the A/D converter, in turn, for sampling.

Two methods are used for multiplexing purposes (Figs. 13.6 and 13.7).

13.2.3.1. Method I

In Fig. 13.6 the multiplexer (also called CMOS bipolar analog switches with channel addresses) samples the input variables by connecting them in sequence to the A/D converter. The disadvantage is that the individual signals are scanned at an interval equal to the minimum processing time of the A/D converter and error creeps in due to possible variation of the input signal during the conversion time (aperture time). Let us find the maximum

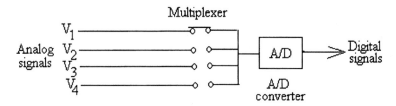

Figure 13.6 Sequential scanning of input signals (no sample-hold).

frequency of the input analog signal that the A/D converter can handle so that conversion accuracy is maintained.

Numerical example. Given the following successive approximation A/D converter specifications:

$T_a = 10$ μsec (aperture or conversion time)

$V_{ac} = \pm 5$ V peak

$m = 16$ bits

Calculate (a) the resolution or least count (RES) and (b) the highest frequency to guarantee accurate conversion of the sample.

Solution:

a.

$$\text{RES} = \frac{5 \text{ V}}{2^{16} - 1} = 152.6 \text{ μV}$$

b. See Fig. 13.8.

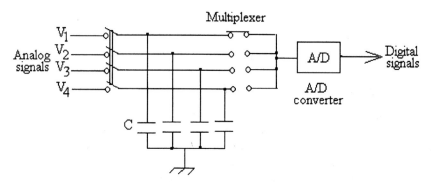

Figure 13.7 Simultaneous acquisition of input signals (with sample-hold circuit).

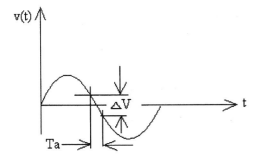

Figure 13.8 Sinusoidal input signal to A/D converter.

Note that the input analog signal should not change by more than one-half the least significant bit (LSB) during the A/D conversion time. Let the input signal $v(t)$, be

$$v(t) = V \sin wt$$

Therefore,

$$\frac{d}{dt}[v(t)] = Vw \cos wt$$

$$= Vw \quad \text{(for } wt = 0\text{)}$$

$$\approx \frac{\Delta V}{T_a} \quad \text{(from Fig. 13.8)}$$

If the input change during conversion time is to be less than half the LSB, we have

$$V_w \geq \frac{\text{RES}/2}{10 \ \mu\text{sec}} \geq \frac{76.3 \ \mu V}{10 \ \mu\text{sec}}$$

Solving for frequency ($w = 2\pi f$, $V = 5$), we obtain $f \leq 0.24$.

Thus, the input signal frequency has to be less than 0.24 Hz to facilitate an accurate conversion. This is too low, since we want 50 to 600 Hz fundamental plus higher frequencies). Therefore it is absolutely necessary to use method II, a sample-hold circuit.

13.2.3.2. Method II

In Fig. 13.7 note that all input signals are sampled at the same instant and stored across the capacitor. The A/D conversion is then done successively. The advantage is that the samples of all the input variables refer to the same instant of time. After all stored analog samples have been converted to digital form, the next sampling sequence starts.

Whatever the method of A/D conversion employed, two basic parameters are important.

1. The sampling rate or samples per power frequency cycle
2. The word length in terms of number of bits used to represent the maximum anticipated value of the input signal and the accuracy at low-input analog signals

Let m = maximum number of bits, or worth length. Then

N = decimal number of all bits corresponding to logic 1 in all bit positions

$$= 2^m - 1 \tag{1}$$

For example, what is the decimal value of the binary number 1111, 1111 (FF hex)?

$$1111\ 1111 = 2^7 + 2^6 + 2^5 + 2^4 + 2^3 + 2^2 + 2^1 + 2^0$$
$$= 2^8 - 1 = 255$$

If the analog input signal is +10 V, then FF hex = 1111 1111 = 255 decimal represents +10 V. The least count, or the minimum value that can be measured, is

$$\Delta V = \text{Least count} = \frac{10}{255} \text{ V} \approx 40 \text{ mV}$$

Thus, the error at the lowest signal V_{min} is

$$\frac{\Delta V}{2V_{min}} \tag{2}$$

If at the instant of sampling the analog value lies between two adjacent digital values (i.e., FF − FE), the sample has to be rounded to either FF or FE (i.e., 255 or 254), for which the maximum error becomes either $\pm 0.5\Delta V$ or ΔV.

Let V_{max}, V_{min} = maximum and minimum values of the signal, and ΔV = permissible error at the lowest signal value. Then from Eqs. (1) and (2)

$$N = 2^m - 1 \geq \frac{V_{max}}{(2\ \Delta V)V_{min}}$$

(note for $m = 8$, $N = 255$.) This enables us to calculate the desired length of the data word m in bits. Standard data word lengths are 8, 16 and 32 bits.

Chapter 13

As the word length increases the minimum signal value that can be measured, with reasonable accuracy, increases.

13.2.4. Verification of Data

Verification is checking whether the digital form of the input variable contains an error. If so, then the verification subroutine should correct the value, if possible. Error correction must be based on the redundancy of data in relation to the minimum amount of essential information. Excess information (redundancy) can be determined in one of the following ways:

1. Repeated measurement of the input variable or measurement of additional variables which permit the variable in question to be determined indirectly
2. Utilization of information available on the range within which the variable could be or its relationship with other variables
3. Expanding the information area and forming forbidden zones

The most common causes of errors are

1. Obvious errors due to an open or a short circuit of CT and PTs or failure of the communication link
2. Errors in signal transducers and transmitters due to overload (saturation of CTs, etc.)
3. Bit errors in the communication of digital signals

The method of state estimation, which has been common practice for years in power system control, includes the detection and correction of erroneous data and is becoming established at substation level. The latter now forms the lowest level of power system state estimation. The following are examples of error correction procedures.

Figure 13.9 illustrates a method of verifying the current measurements in three feeders with Kirchhoff's law, namely the magnitude of the currents entering a junction must equal the magnitude of the currents leaving it:

$$I_1 + I_2 = I_3$$

An error exists whenever this condition is not fulfilled.

The second method (Fig. 13.10) uses an excess signal. The secondary voltages V_a, V_b, V_c and V_0 are measured and, assuming a standard VT ratio, the following condition must be satisfied:

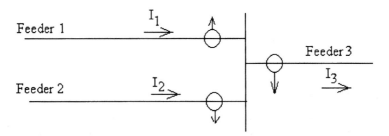

Figure 13.9 Signal verification by Kirchhoff's law.

$$V_a + V_b + V_c = V_\Delta = 3V_0$$

An error exists whenever this condition is not fulfilled.

The formation of forbidden zones involves establishment of codes when estimating the transmission errors of the individual signal bits. The simplest code for this purpose uses a parity bit; in other words, a bit is added to every word to make the total number of bits with value 1 even. The parity bits for 0 to 9 are

Number	Word	Parity bit
0	0000	0
1	0001	1
2	0010	1
3	0011	0
4	0100	1
5	0101	0
6	0110	0
7	0111	1
8	1000	1
9	1001	0

This code can detect single word errors but is powerless to detect double errors.

A much more effective, but complicated, code is called the *same index code*. It relies on the fact that the same number of 1's must exist in every word. Assuming a word has m bits and the number of 1's is k, then

$$\text{No. digital states} = \frac{n!}{k!}(n-k)$$

Chapter 13

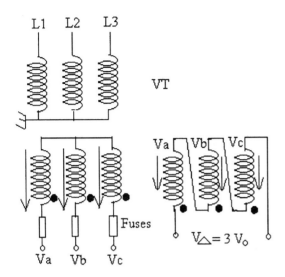

Figure 13.10 Signal verification by excess information.

The number of digital states which can be expressed by a noncoded word is $2^m - 1$.

The 2-out-of-5 code is frequently used in practice; i.e., the number of 1's in a 5-bit word is fixed at 2. The words for the first 10 digits are

Number	Word
0	10000
1	00011
2	00101
3	00110
4	01001
5	01010
6	01100
7	10001
8	10010
9	10100

This code detects most multiple errors in addition to single errors. Note, however, that the forbidden zone is twice as large as the permissible zone. This is why the possibilities of writing many state numbers are reduced.

13.3. COMPUTER PROTECTION

13.3.1. Advantages of Computer Relaying

13.3.1.1. Flexibility

A single general-purpose hardware base can be used to perform a variety of protection and control functions with change of programming only. Several of these functions can reside in the processor ROM simultaneously and can be called upon when needed. Similarly, drastic alterations or additions to protection logic can be made in the field, with little or no hardware replacement.

13.3.1.2. Adaptive Capability

A processor can be programmed to change its behavior automatically, depending on external circumstances which change with time. The basis for change can be either local information available directly to the processor, such as prefault load flow in the protected apparatus; or the change can initiate from an external source of intelligence, such as a substation operator or a data link from a central system control computer. The change may be only in a specific setting, or a new protection routine can be selected when needed.

13.3.1.3. Data Interface Access

General-purpose digital computer systems can always be equipped with input/output ports through which data and control commands can be exchanged. For example, the sequence of software events which occurs in the processor in response to a fault can be stored during the fault and output onto a data link afterward. A control computer or operator is provided with detailed information on fault specifics and the resulting software action. In turn, data can be input to the processor memory to improve or modify the relay characteristics, etc.

13.3.1.4. Self-Checking Ability

Conventional relays are idle for essentially their entire lives—faults are present for, perhaps, a few seconds out of 20 or 30 years of relay life. At other times the ability of the equipment to do its job is in doubt and must be confirmed by periodic testing. A digital processor, on the other hand, is by nature a dynamic device, and most hardware failures are flagged by a processor stop as soon as they occur. Beyond this, specific programs can be executed during no-fault periods which test the processing hardware, the integrity of the program memory, the calibration of the A/D interface, and much more.

13.3.1.5. Economic Benefits

While the cost of conventional relays steadily increases, the cost of relays using digital processors, suitable for apparatus protection, is plummeting. This trend is likely to continue. The prospective user should remember, though, that software development costs are not declining and will dominate hardware costs, unless identical systems are built.

13.3.1.6. Logical and Mathematical Abilities

While the conventional relay designer is somewhat constrained by the characteristics and limitations of electromechanical or solid-state relays, the relaying programmer is free to provide almost any function within the limits of his or her imagination or understanding. Specific protection problems can be broken down into fine detail and handled separately.

Asea Brown Boveri and Siemens list the following features of their numerical relays:

1. Compact design, few hardware units
2. Selectable protection functions
3. Many applications
4. Setting menu-assisted with personal computer or any other man-machine communication (MMC)
5. Fully numerical signal processing
6. Continuous self-monitoring by hardware
7. Cyclically executed testing routines, mostly by software
8. Setting parameters and recording the settings
9. Display of measured values
10. Display of events, their acknowledgment and printout
11. Disturbance recording
12. Self-documentation
13. Long-term stability
14. Serial ports for communication

A simple example related to distance protection of transmission lines is the ease with which relay characteristic shapes in the R-X diagram can be provided. In addition, measurement problems can be stated as mathematical equations and directly implemented.

Many investigators of computer relaying have recognized and proposed possible equations which can serve as a basis for distance-type protection of EHV lines.

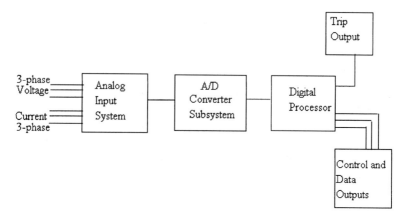

Figure 13.11 Simplified hardware configuration.

Although a broad spectrum of arrangements is possible for application of computers to relaying, the most challenging part is EHV line protection. Figure 13.11 shows a simplified hardware configuration for distance protection.

An analog input subsystem accepts three-phase ac quantities from power system transducers, such as conventional CTs and PTs. All of the quantities are sampled, simultaneously, at a uniform rate from 16 to 32 times per ac cycle, converted to digital form, and transferred to the digital processor. The processor stores, organizes, and makes decisions based on the values of the samples. Its prime purpose is to output a breaker-tripping command when a fault occurs on the protected line. Other secondary control and data outputs may also be provided.

Stored in the processor ROM is an elaborate program which implements the line protection sensing and logic. At the core of the relaying program is an algorithm or equation which operates on the incoming raw data samples to provide meaningful indicators of fault presence or location.

13.3.2. Progress to Date

The first widely recognized milestone in computer relaying is the 1968 paper by Rockfeller. In 1970 Mann and Morrison described the derivation and testing of programs for line protection. Many other investigators saw that the economic trend favored the digital approach for the future, and proposed new, ingenious algorithms and related processing techniques. Ramamoorty, Phadke, Carr, Jackson, McLaren and others proposed various forms of Fourier analysis, sometimes expressed as a waveform correlation.

Chapter 13

The key departure taken by these authors is that the fault voltage and current waveforms are no longer assumed to be sinusoidal. Components other than 50- or 60-Hz fundamental are expected; the correlation process acts like a notch filter. Horton, Hope and many others described the notion of correlation notch filtering using waveforms other than sinusoids, notably square waves.

Viewing the problem in yet a different way, McInnes and Morrison considered a simple R-L lumped parameter model of a transmission line. They developed, from the differential equations describing that model, an algorithm which yields R and L values from the relaying point to the fault directly. Decaying dc offsets in the current are included in the solution and do not cause errors as in some other simpler approaches. Ranjbar and Cory suggested modification in which limits of integration are chosen to suppress specific undesired harmonics, not accounted for in the lumped R-L model. Gilbert and others proposed methods which can loosely be categorized as curve fits.

Such methods represent rather complete sampling techniques which have been proposed for postfault steady-state impedance measurement used in line protection. At the moment distance relays are being proposed on traveling-wave techniques, etc.

13.3.3. Listing of Line Protection Algorithms

The suggested algorithms are

1. Fourier analysis—one-cycle window
2. Fourier analysis—short data window
3. Walsh analysis
4. Sample-and-derivative algorithm
5. First- and second-derivative algorithms
6. Solution of differential equation
7. Solution of differential equation with selected limits of integration
8. Sinusoidal curve fit
9. Sinusoidal curve fit—short window
    a. Without prefiltering
    b. With digital prefiltering
10. Least curve fit

13.3.3.1. Fourier Analysis—One-Cycle Window

The one-cycle window Fourier analysis is the most widely accepted technique for almost all relaying purposes and measurement and, therefore, is discussed in detail. The incoming ac data samples for one cycle are correlated with the stored samples of reference fundamental sine and cosine waves to extract the complex value of the fundamental component in rectangular form.

The general expressions for the sine and cosine components of voltage at a sample point k are

$$V_s = \frac{1}{N}\left[2\sum_{l=1}^{N-1} v_{k-N+l} \sin \frac{2\pi}{N}\right]$$

$$V_c = \frac{1}{N}\left[2\sum_{l=1}^{N-1} v_{k-N+l} \cos \frac{2\pi}{N}\right]$$

$v_i = i$th voltage sample

N = number of samples taken per fundamental cycle

Similar expressions are evaluated for current components I_s and I_c; the four results can be used to generate the phasor impedance value in polar or rectangular form. In polar form,

$$\text{modulus of } Z = \left|\frac{V_s^2 + V_c^2}{I_s^2 + I_c^2}\right|^{1/2}$$

$$\theta_k = \arctan \frac{I_s}{I_c} - \arctan \frac{V_s}{V_c}$$

Implicit in Fourier analysis is

1. The drastic filtering of data: The dc offset and all harmonics up to $N/2$ are completely filtered out.
2. Slow response: the output responds slowly but accurately to badly distorted postfault waveform.
3. High processing time: the processing time is high due to multiplications, but can be reduced by parallel processors, a math coprocessor or a digital signal processor.

This Fourier technique is being universally employed in relaying, and the hardware costs are going down. The frequency response is excellent, as shown in Fig. 13.12. There is complete filtering (null points) of dc offset and harmonics up to half the sampling rate ($N/2$ by Shannon's theorem). The antialiasing analog filter needs to filter out harmonics above $N/2$. The slopes between the null points are called leakages and are due to the possible drift in power frequency. The solution is to have the sampling frequency as a multiple of drifting power frequency by a phase-locked loop.

Chapter 13

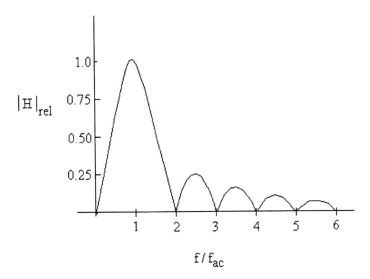

Figure 13.12 Frequency response of Fourier analysis—one-cycle window.

13.3.3.2. Walsh Analysis

The Walsh algorithm is closely related to Fourier analysis. The orthogonal functions correlated with the fault waveform, however, are not fundamental sine and cosine waves, but odd and even square waves. Real-time computation is simplified since the reference square waves are denoted simply as + or −. However, some of this benefit is mitigated by the need to extract harmonics of the square-wave components along with the fundamental so that the desired fundamental sinusoidal component can be reconstructed.

Similar to Fourier analysis, source data filtering is drastic, and the resultant output is thus heavily damped. Accuracy is good even for distorted input waveforms. Figure 13.13 shows the frequency response for a sampling rate of eight samples per cycle.

13.3.3.3. Sample-and-Derivative Method

The sample-and-derivative algorithm was first proposed for the characterization of the sinusoidal waveform from asynchronous data samples. Let

$$v = V_m \sin(wt + \theta_v)$$

Differentiating gives

$$v' = \frac{dv}{dt} = V_m w \cos(wt + \theta_v)$$

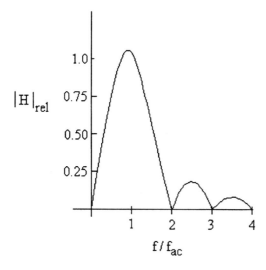

Figure 13.13 Frequency response of Walsh analysis.

These equations are solved to yield the peak magnitude and phase in terms of v and v' at an arbitrary sampling instant.

$$\text{Modulus of } V = |v_k^2 + (v_k')^2|^{1/2}$$

$$\theta = \arctan \left| \frac{v_k}{v_k'} \right|$$

The slope v' is approximated by the difference between two samples straddling the central sample point:

$$v_k' = \frac{1}{hw} |v_{k+1} - v_{k-1}|$$

where h = sampling interval in seconds
w = fundamental angular frequency = $2\pi f$

With a window of just three close samples, the algorithm responds quickly to sudden changes in the input faulted voltage waveform. Dc offsets, however, upset the expected relationship between the sample value and its first derivative by changing only the former. Similarly, harmonics can produce large errors. Figure 13.14 shows the frequency response and its inadequacy. The response plot shows a peak at the third harmonic (instead of fundamental) with no nulls at any multiples of the fundamental.

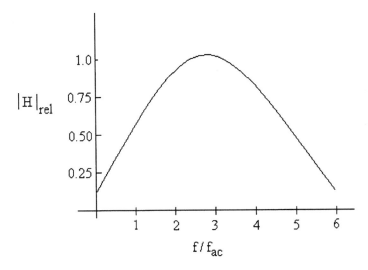

Figure 13.14 Frequency of sample-and-derivative algorithm.

13.3.3.4. Other Methods

Without writing the equations, we give the frequency responses of other methods.

Fourier analysis with short data window. The basis of this method is the same as Fourier, but the integration period or data window is shortened to half a cycle. The frequency response plot (Fig. 13.15), when compared to a full-cycle window, shows reduced effectiveness in coping with dc offsets and even harmonics.

First- and second-derivative calculations. The frequency response plot (Fig. 13.16) shows that the dc offset is filtered out, but all even and odd harmonics are present and amplified.

Solution of differential equation. The frequency response plot (Fig. 13.17) shows that there is considerable attenuation of the high-frequency components, but the dc offset is fully present.

Solution of differential equation with selected limits. The frequency response plot (Fig. 13.18) shows improved suppression of high odd and even harmonics but not the dc offset.

Sinusoidal curve fit. This algorithm is highly susceptible to high-frequency components. Note the peak at the third harmonic (Fig. 13.19).

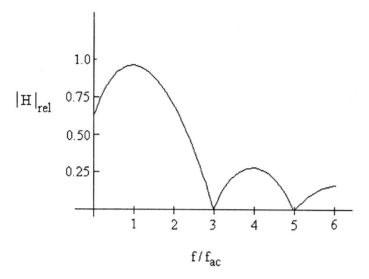

Figure 13.15 Frequency response of Fourier analysis with short data window.

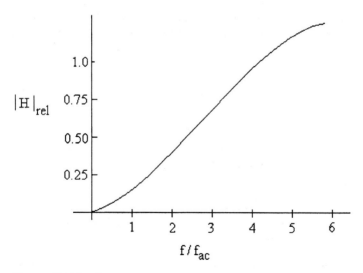

Figure 13.16 Frequency response of first- and second-derivative algorithms.

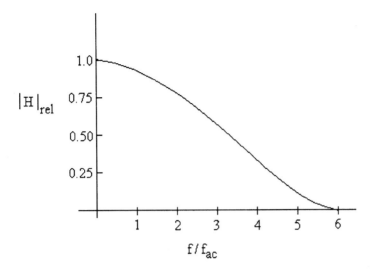

Figure 13.17 Frequency response of solution of differential equation.

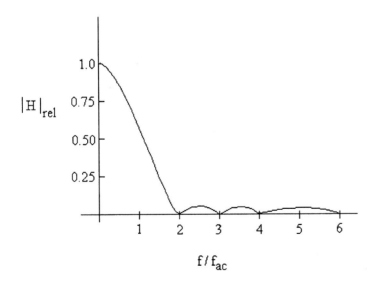

Figure 13.18 Solution of differential equation with selected limits of integration.

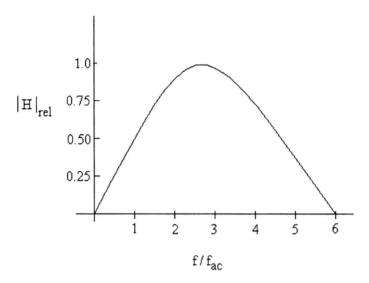

Figure 13.19 Frequency response of sinusoidal curve fit.

Sinusoidal curve fit with short data window. Figure 13.20 exhibits considerable error at all frequencies. Makino and Miki proposed additional digital prefiltering, which gives improved frequency response (Fig. 13.20). It has been reported that with additional prefiltering the processing time increases drastically.

Least-square fit. The frequency response (Fig. 13.21) shows a peak near, but not exactly at, the fundamental frequency, with lesser response to high frequencies.

13.4. LOGICAL STRUCTURES FOR DIGITAL PROTECTION

13.4.1. Introduction

One of the most striking advantages of programmable digital relays is the ease with which logical conditions and relationships can be checked. As a consequence of the large volume of data and logical criteria which modern digital devices can process, it has been possible to expand the information exchanged between the protected unit and protection devices, to increase the speed and accuracy of fault detection and location, and enabled the protection to take on new tasks.

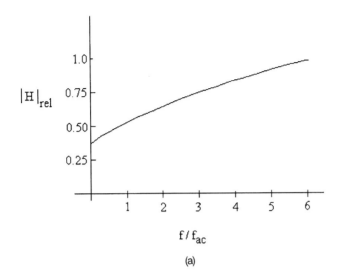

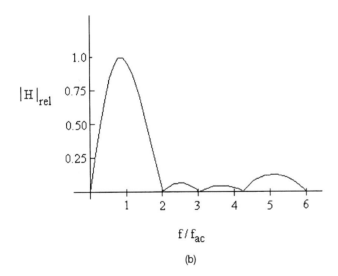

Figure 13.20 (a) Sinusoidal curve fit—short data window (without prefiltering); (b) sinusoidal curve fit—short data window (with digital prefiltering).

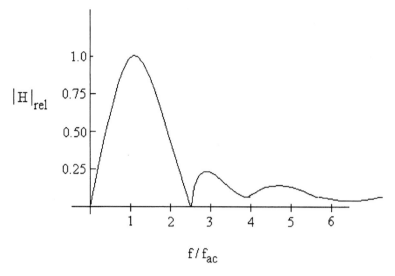

Figure 13.21 Frequency response of least square fit.

The logical structure of complex protection devices and systems can be divided into the following functional areas:

1. Initial fault detection
2. Determination of whether the fault is in the assigned zone of protection
3. Adjustment of zone protection to suit the state of the system, such as reduction of the first zone of a distance relay after the first autoreclosure attempt
4. Selection of faulted phase or phases
5. Exchange of data with other protection devices or systems, such as with a protection relay at the opposite end of line
6. Tripping control and communication with operating personnel
7. Automatic self-supervision, thereby eliminating the periodic maintenance

As an example, the following section is only concerned with fault occurrence and its selective decision, particularly for a distance relay.

13.4.2. Logical Structure for Determining the Operating Characteristics of a Distance Relay

The logical structure of a distance device is especially complicated, because it can include alternative configurations in addition to the functional areas listed earlier. Only the functional area directly concerning fulfillment or nonfulfillment of the condition for operation of a digital distance relay is described. It involves determining whether the measured resistance and reactance of the protected line lie inside or outside the operating characteristic shown in Fig. 13.22. The diagram assumes a two-zone relay with the first zone undelayed (i.e., $t = 0$) and the second zone delayed by time T_2.

One possible logical structure of a distance protection device is shown in Fig. 13.23. Block 1 determines whether the measured resistance R is to the right of the straight line $R = -R_d$. Similarly block 2 determines whether the measured reactance X_s is above the straight line OA with slope $-h_a$. If these conditions are fulfilled, block 3 checks that both resistance and reactance are below the first zone limit line X_b to B, with slope $-h_b$. If this condition is fulfilled, block 4 checks whether the resistance R_s is less than the first zone limit R_b. If this condition is also fulfilled, then the fault must lie in the first zone and the distance relay trips.

Assuming the condition determined by block 3 is not fulfilled, block 5 checks whether the reactance is below the limit line X_c to C for the second zone. Should the condition be fulfilled, block 6 determines whether the resistance R_s is to the left of the second zone limit line R_c. As soon as all conditions for the second zone are fulfilled, the time T_2 in block 7 starts to run and tripping takes place in block 8, when it expires.

As pointed out earlier, this is just one of the many possible structures which can be expanded to measure more zones or to include other conditions. The slopes h_a, h_b and h_c should be chosen to minimize the multiplications.

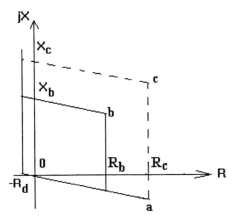

Figure 13.22 Typical operating characteristics of QDR.

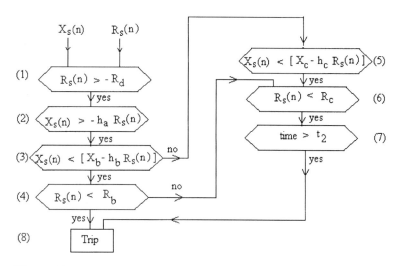

Figure 13.23 Typical logical structure for a QDR.

13.5. DESIGN OF DIGITAL PROTECTION AND CONTROL DEVICES

The most general arrangement for the hardware for integrated protection and control devices can be seen from Fig. 13.24. The analog input variables (current and voltage) are electrically insulated from the internal circuits by input transformers, with grounded screens between primaries and secondaries. The antialiasing filter is basically a low-pass filter whose cutoff frequency depends on the sampling rate chosen for the A/D converter, which finally determines the frequency band to be transmitted. Since the multiplexer scans the input variables sequentially, it is equipped with sample-hold capacitors. The instantaneous values held across the capacitor are now converted by the A/D converter sequentially one by one. All these processes are controlled by the front-end CPU, as shown in the figure.

The binary number now passes via a dual-port memory (DPM) and a digital filter to the main CPU, which runs the programmed protection algorithm. Depending on the scope of the protection tasks to be performed, several main CPUs may operate in parallel on the bus. The results are then communicated to a logic CPU via a further DPM. The purpose of DPMs is to separate as much as possible functions of different CPUs.

From the results supplied by the main CPUs and from digital inputs, interlocks and settings, the logic processor derives the commands needed for tripping, signaling, etc. The command signals go to the input/output units

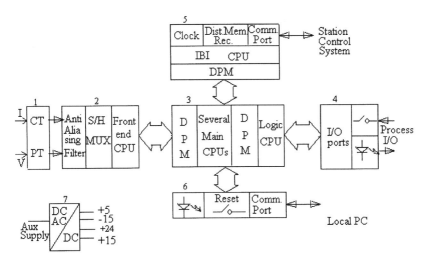

Figure 13.24 Block diagram of an integrated protection and control device: (1) input transformer device; (2) A/D converter; (3) main processor unit; (4) binary I/O unit; (5) communication unit; (6) connecting unit; (7) auxiliary supply unit.

and ports, which transmit them to the tripping CBs and other functions. As with input analog signals, the output signals are also isolated by high-speed reed relays or optocouplers.

Interfaces for the local control unit, consisting of a personal computer and a keypad, also called the object bus interface (OBI), are also connected to the main CPUs via a DPM. The OBI attaches a time marker to all the data requested from the device. The values of analog and digital variables stored in the device before, during and after disturbance can be assessed by the logic control unit when the device is operating in a stand-alone mode or by the OBI when it is connected to a station control system.

The power supply unit provides the electrical insulation between the device and the station battery and generates all the auxiliary supplies needed. The function of the power supply unit, as well as all other parts of the device, is continuously supervised. An example of a self-supervision system is given in Fig. 13.25.

The following are supervised in addition to the auxiliary supplies and the input currents and voltages:

1. A/D converter by continuously switching reference voltage
2. Memories by write/read test cycles
3. Data communication by error codes
4. Processors by watchdogs

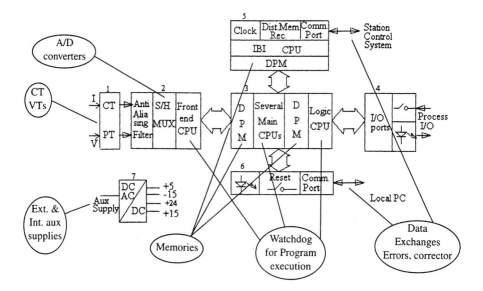

Figure 13.25 Continuous self-supervision.

13.5.1. Signal Flow

The flow of data and signals for the example in Figs. 13.24 and 13.25 can be seen in Fig. 13.26. Input transformers electrically insulate the analog input variables from the initial circuits before shunts convert the currents into proportional voltages. The signals pass in parallel via the (antialiasing) analog low-pass filter, amplifier and sample-hold memory to the multiplexer. The multiplexer converts them in serial form for the A/D converter. In the A/D converter all data are transferred and processed serially.

The signals are now processed in accordance with the protection algorithms in the main CPUs, which then pass the results to the logical processor, where they are related to other input signals and settings being transformed into commands for the process. Data for evaluating disturbances and control data needed by the station control system are taken directly from the main CPUs.

The Asea Brown Boveri has its own general arrangement for the numerical protection as shown in Fig. 13.27. It basically uses eight digital signal processors for

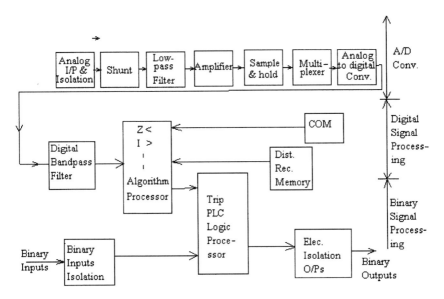

Figure 13.26 Flow of data and signals in an integrated protection and control device.

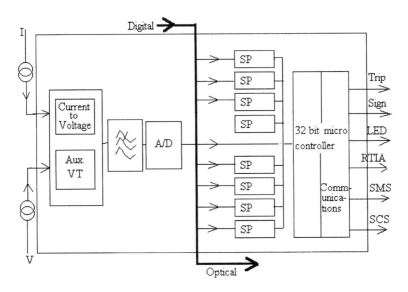

Figure 13.27 Asea Brown Boveri basic block diagram for numerical relay.

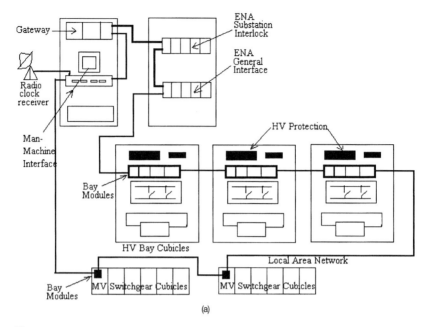

Figure 13.28 (a) General Electric company integrated digital substation control system: the General Electric Company's integrated digital substation; (b) numerical relay by GEC; (c) a numerical relay by Siemens.

1. (SP 1) calculation of impedances for all L-G faults
2. (SP 2) calculation of impedances for all phase faults
3. (SP 3) directional features for all six measuring units
4. (SP 4) phase selection, power swing detection, weak-end feed, etc.
5. (SP 5) earth-fault OC protection
6. (SP 6) fault locator function
7. (SP 7, 8) two additional distance zones, if required
8. (RTIA) man-machine interface
9. (SMS) substation monitoring system
10. (SCS) substation control system

The integrated digital substation arrangement suggested by the General Electric Company is shown in Fig. 13.28a. It provides the following functions:

1. Plant and network protection
2. Archiving and real-time recording
3. Automatic system configuration to maintain continuity of supply

Chapter 13

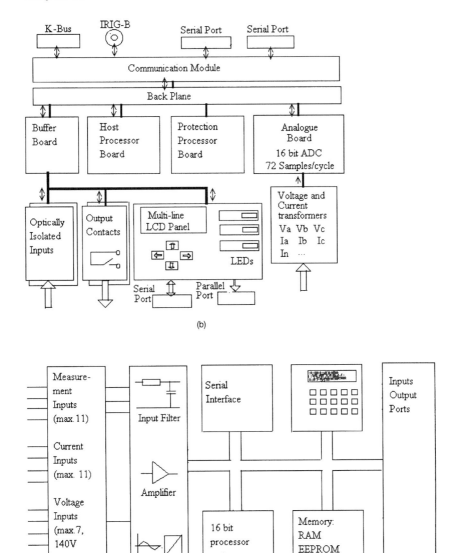

Figure 13.28 Continued

4. System synchronization
5. Sequence of events recording
6. Local and remote alarm annunciation
7. Local control point at substation level
8. Remote control and supervision

A typical hardware configuration for a numerical high-speed distance protection scheme by General Electric is shown in Fig. 13.28b, whereas the Siemens system is shown in Fig. 13.28c.

13.5.2. System Architecture

The system architecture for the example in Fig. 13.24 is given in Fig. 13.29. Random access memory (RAM) units are used for the main memory, which permits intermediate data to be written, read or changed while processing. The application program and algorithms are factory programmed in ROM. The instructions and data coded in ROM can be read only and are not destroyed by failure of the auxiliary dc supply to processors.

Settings and information concerning protection configuration are stored in EEPROMs. This data can be read on entering the password. On failure of the dc supply, the data remains intact.

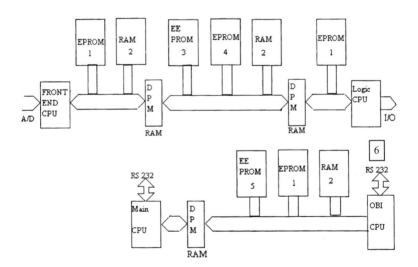

Figure 13.29 System architecture: (1) program memory; (2) main memory; (3) parameter setting; (4) program memory (settings, etc.); (5) configuration memory; (6) dist. rec. memory (buffered).

13.5.3. Operator Control

The operator normally controls the device by a PC connected to the interface. Provision is also made for automatic control by a central bay computer connected to the station control system. This facilitates menu-guided dialogue with the device and more sophisticated displays. Many times additional LED displays continuously display the most important information, such as the currents and voltages measured, real and reactive power calculated, frequency, settings and relay response.

The processor also stores values of analog and digital signals before, during and after a disturbance, with common time markers so that they can be assessed. This data can also be outputted on the printer in the form of system line diagrams to aid fault diagnosis.

A list of settings can be produced with time markers to every signal. New sets of settings can be prepared off-line on the PC and then downloaded to the device either locally or via a communication link. Resetting also includes necessary changes in signaling and tripping logic.

13.6. ADAPTIVE PROTECTION SYSTEMS

Adaptive protection systems permit protective functions to be adapted automatically, without manual intervention, in real time to changing system conditions. Some examples of adaptive protection are

1. Adjusting the slope of biased differential relaying in relation to tap changes in transformers and increased through-fault levels.
2. Changing the zone 2 and zone 3 settings of the distance scheme according to the status of the parallel line. Note that the parallel line is an infeed for zone 2 and zone 3. In no case should zone 3 enter the distribution level.
3. Intelligent load-shedding program for the underfrequency relay and rate of change of frequency relay according to the actual mismatch of active power and the loads the various feeders are carrying at the instant of islanding.
4. Modifying the autoreclosure program according to whether the parallel line is in or out.

13.7. EXPERT SYSTEMS

A recent paper, ''Potential Applications of Expert Systems to Power system Protection,'' by the IEEE working group (IEEE *Transactions on Power De-*

livery, 9(2) 1994) is worth studying for solving current protection problems by expert system methodology.

An *expert system* (ES) is a computer program that uses knowledge and inference procedures to solve problems that are ordinarily solved through human expertise. A knowledge engineer obtains the necessary knowledge from a human expert and puts it into a knowledge base. This knowledge, together with the inference procedures used, can be thought of as a model of an expert in the specialized field or domain of interest. The basic components of an expert system are

>Knowledge base
>Inference procedures
>Data base
>User interface

The problem-solving control strategies employed are forward chaining, backward chaining, or both. A forward-chaining inferencing control strategy is intrinsic to planning and configuration problems. A backward-chaining strategy is well suited to diagnostics. The following protection problems can be solved by expert systems:

1. Protection equipment failure and event diagnostics
2. Protection selection, setting and coordination
3. Fault identification and location
4. Service restoration and remedial action
5. Distribution feeder protection

Briefly, the problem-solving procedure is as follows:

1. The expert system analyzes each and every past case, including the present data, and arrives at a decision, in a broad sense, for the changes in protection and control settings and procedures.
2. The analysis takes into account the trends in data changes and their mutual relationship. Only that area is examined which is likely to influence the decisions to be made.
3. Where the data are incomplete, the expert system arrives at its decision on the basis of its own experience and convictions in consideration of the probability of the events and estimation of wrong decisions.

So far, the expert system is still in the formulating stage, due to the difficulty in developing the software and the strict requirements regarding the speed with which protection decisions have to be made. Even then researchers are forecasting the development of expert systems.

13.8. PROBLEMS AND EXERCISES

1. Consider an 8-bit A/D converter. If the A/D converter generates a $[1111\ 1111]_2$ or $[FF]_{hex}$ word for an input analog voltage of $+10$ V, what are the voltages corresponding to the following data words FE, FD, 02 and 01 in hex? Draw conclusions, if any.

2. An m-bit word is required to measure currents from $+1.0$ to $+80.0$ A. The accuracy required to measure at the lowest current of 1.0 A is 1.0%. Find the minimum word length m.

3. Evaluate

$$I = \int_{x=-\pi}^{x=\pi} \sin nx \sin x \, dx$$

   for $n = 1$ and for $n \neq 1$.

4. Prove that the first-order Walsh function (i.e., ± 1 square wave)

$$f(x) = \begin{cases} -1.0 & -\pi \leq x < 0 \\ +1.0 & 0 \leq x \leq \pi \end{cases}$$

   can be expanded as

$$f(x) = \sin x + \frac{1}{3}\sin 3x + \frac{1}{5}\sin 5x + \cdots$$

5. Plot the frequency response of the filter

$$\frac{v_o(S)}{V_i(S)} = \frac{1}{S^3 + 2S^2 + 2S + 1}$$

6. The specifications of an A/D/ converter are
   a. $T_{conversion} = 50\ \mu sec$
   b. ± 10 V bipolar sinusoidal input
   c. 12 bits

   Find the resolution of the converter and the highest frequency of the input sinusoidal voltage, without a sample-hold capacitor, that the A/D converter can convert within $\pm 1/2$-bit accuracy.

Bibliography

General Reading

GI Atabekof. Relay Protection of H.V. Networks. Pergamon Press, 1960.

Bibliography of Relay Literature, IEEE Committee Report. The latest of a series of classified lists of power system relaying references, begun in 1927, is presented in the following table.

Volume	No.	Year	Pages
1927–1939	60	1941	1435–1447
1940–1943	63	1944	705–709
1944–1946	67,Pt I	1948	24–27
1947–1949	70,Pt I	1951	247–250
1950–1952	74,Pt III	1955	45–48
1953–1954	76,Pt III	1957	126–129
1955–1956	78,Pt III	1959	78–81
1957–1958	79,Pt III	1960	39–42
1959–1960	81,Pt III	1962	109–112
1961–1964	PAS-85,10	1966	1044–1053
1965–1966	PAS-88, 3	1969	244–250
1967–1969	PAS-90,5	1971	1982–1988
1970–1971	PAS-92,3	1973	1132–1140
1972–1973	PAS-94,6	1975	2033–2041
1974–1975	PAS-97,3	1978	789–801
1976–1977	PAS-90,1	1980	99–107
1978–1979	PAS-100,5	1981	2407–2415
1980–1981	PAS-102,4	1983	1014–1024
1982–1983	PAS-104,5	1985	1189–1197
1984–1985	PWRD-2,2	1987	349–358
1986–1987	PWRD-4,3	1989	1649–1659
1988–1989	PWRD-6,4	1991	1409–1422
1990	PWRD-7,1	1992	173–181
1991	PWRD-8,3	1993	955–961

JL Bleckburn, Editor. Applied Protective Relaying. Newark, NJ: Westinghouse Electric Corporation, 1976.
Central Electricity Council. Power System Protection. Vols I, II, III. London: Macdonald, 1969.
V Cook. Analysis of Distance Protection. Letchworth, Hertfordshire, UK: Research Studies Press, 1985.
DF Elliot et al. Fast Transformers: Algorithms, Analysis, Applications. Academic Press, 1982.
SH Horowitz, Editor. Protective Relaying for Power Systems I. IEEE Press, 1980.
SH Horowitz, Editor. Protective Relaying for Power Systems II. IEEE Press, 1991.
SH Horowitz et al. Power System Relaying. New York: Wiley, 1991.
EW Kimbark. Power System Stability. Vols I, II, III. New York: Wiley, 1956.
CR Mason. The Art and Science of Protective Relaying. New York: Wiley, 1956.
P Mathews. Protective Current Transformers and Circuits. Chapman and Hall, 1955.
AG Phadke et al. Computer Relaying for Power Systems. Letchworth, Hertfordshire, UK: Research Studies Press, 1988.
Protection Engineering Department. GEC Measurements. Stafford, UK, 1987.
BD Russel et al. Power System Control and Protection. Academic Press, 1978.
AR van C Warrington. Protective Relays (Their Theory and Practice). Vols I, II. London: Chapman & Hall, 1977.

Chapter 1

AIEE Relay Committee Report. Recent practices and trends in protective relaying. Vol 78, Part IIIB 1959, pp 1759–1777.
AIEE Relay Committee Report. A review of back-up relaying practice. Vol 72, Part III, 1953, pp 137–141.
AIEE Committee Report. Power system fault control. AIEE Trans 70:410–417, 1951.
EL Harder et al. Principles and practices of relaying in United States. AIEE Trans 67(Part II):1005–1022, 1948.
AR van C Warrington. A consideration of theory of relays. Gen Ele Rev 43(9): 370–373, 1940.

Chapter 2

AIEE Relaying Committee. Supplement to recent practices and trends in protective relaying. IEEE Press, Protective Relaying for Power Systems, 1980, pp 38–42.
AIEE Relay Committee Report. A review of back-up relaying practice. Trans AIEE 72(Part III):137–141, 1953.
LF Kennedy et al. An appraisal of remote and local back-up relaying. IEEE Press, Protective Relaying for Power Systems. 1980, pp 42–49.
WA Lewis et al. Fundamental basis for distance relaying on three-phase systems. Trans AIEE 66:694–709, 1947.

Bibliography

Chapter 3

AIEE Technical Committee. Technical Paper 52-53. Progress report on coordination of protection of distribution circuits. 1949.
AIEE Committee. Sensitive ground protection. AIEE Trans 69(Part I):475–476, 1950.
JL Blackburn. Ground relay polarization. AIEE Trans 71:1088–1093, 1952.
JL Blackburn. Ground fault relay protection of transmission lines. AIEE Trans 71: 685–691, 1952.
FP Brightman. Selecting protective device settings for industrial plants. AIEE Trans 71(Part II):203–211, 1952.
D Dalasta et al. An improved static overcurrent relay. Trans IEEE (PAS) 68:705–716, 1963.
IEEE Power System Committee Report. Relaying to detect ground faults on transmission, subtransmission and distribution line. IEEE Trans (PAS) 85:524–532, 1966.
MJ Lantz. Effect of fault resistance on ground-fault current. AIEE Trans 72(Part III):1018–1019, 1953.
M Lathrop et al. Protective relaying on industrial power systems. AIEE Trans 70(Part II):1344, 1951.
M Ramamoorthy. Application of digital computers to power system protection. JIE (India) 52(Pt. e15):235–238, 1972.
WK Sonneman. A new single-phase-to-ground fault detecting relay. AIEE Trans 61: 677–680, 1942.
WK Sonnemann. A study of directional element connections for phase relays. AIEE Trans 69(Part II):1438–1450, 1950.
WK Sonneman. A new overcurrent relay with adjustable characteristics. Trans AIEE 72(Part III):23–27, 1953.

Chapter 4

C Adamson et al. Power system protection with particular reference to the application of junction transistors to distance relays. Proc. IEE (London) 103(Part A):379, 1956.
C Adamson et al. A dual comparator MHO type distance relays utilizing transistors. Proc IEE (London), 103(Part A):509, 1956.
AIEE Committee. Interim report on application and out-of-step protection. AIEE Trans 62:567–573, 1943.
JL Blackburn. Ground relay polarization. AIEE Trans (PAS) 71:1088–1093, 1952.
L Calhoun et al. Zone packaged ground fault distance relay II. IEEE Trans (PAS) 85:1128–1134, 1966.
WA Elmore. Zero sequence mutual effects on ground distance relays and fault locators. 45th Annual Texas A & M Protective Relay Conference, 1992.
SL Goldsborough. A new distance ground relay. AIEE Trans 67(Part II):1442–1446, 1948.

T Gudmundsson et al. Numerical protection for power transmission systems. ABB Rev 3–8, 1992.

H Gutmann. Behaviour of reactance relays with short circuits fed from both ends. ETZ 514, 1940.

FL Hamilton. The application of transductors as relays in protective gear. Proc. IEEE (London) 99(Part II):297, 1952.

RM Hithinson. The MHO distance relay. AIEE Trans 65:353–360, 1946.

IEEE Power Systems Relaying Committee. EHV protection problems. IEEE Trans (PAS) 100:2399–2405, 1981.

IEEE Working Group. Arc deionising times on high-speed three pole reclosing. IEEE Trans Special Suppl, 1963.

AT Johns et al. Hierarchical protection of transmission systems. 5th International Conference on Developments in Power System Protection, 1993, pp 123–126.

WA Lewis et al. Fundamental basis for distance relaying on three-phase systems. AIEE Trans 66:697–708, 1947.

S Liberman et al. Ultra high-speed relay for EHV/UHV lines. IEEE Trans (PAS), 97:2104–2112, 1978.

PG Maclaren et al. Open relaying systems—a new philosophy. 5th International Conference on Development in Power System Protection, 1993, pp 95–98.

RH Macpherson et al. Electronic protective relays. AIEE Trans 67(Part II):1702–1279, 1948.

WC Morris. One slip cycle out-of-step relay equipment. Trans AIEE 68(Part II):1246–1248, 1949.

IF Morrison et al. Digital calculation of impedance for transmission line protection. IEEE Trans 90(1):270–279, 1971.

YG Paithankar. Fast (1-shift) orthogonal functions for extraction of fundamental frequency component for computer relaying. EPSR 14:233–236, 1988.

J Roberts et al. Z = V/I does not make a distance relay. 20th Annual Western Protective Relay Conference, 1993.

GD Rockfeller. Zone packed ground distance relay I. IEEE Trans (PAS) 85:1021–1044, 1966.

GD Rockfeller. A modern view of out-of-step relaying. IEEE Conference paper, 31 CP 66-34.

GD Rockfeller. Fault protection with digital computer. IEEE Trans 88:438, 1969.

HJ Sutton. The application of relaying on an EHV system. IEEE Trans (PAS) 86:408–415, 1967.

HR Vaughan. Out-of-step blocking and selective tripping with impedance relays. AIEE Trans, 58:637–645, 1939.

CL Wagner. Islanding problems for non-utility generation. 45th Annual Texas A & M Protective Relay Conference, 1992.

AR van C Warrington. Protective relaying for long transmission lines. AIEE Trans 62:261–268, 1943.

AR van C Warrington. Interesting facts about power arcs. GEC Relaying News, 1941.

AR van C Warrington. Graphical method for estimating the performance of distance relays during faults and power swings. AIEE Trans 68:608–620, 1949.

LM Wedephol. Polarised MHO relay. Proc. IEE (London) 122:525–535, 1965.

Chapter 5

MG Adamiak. Communication requirements for protection and control in the 1990's. Texas A&M Protective Relaying Conference, 1991.
F Calero et al. Current differential and phase comparison relaying schemes. Trans CEA E&O Div, 32:paper no. 93-sp-96, 1993.
RC Cheek et al. Considerations in selecting a carrier relaying system. AIEE Trans, 71:10–15, 1952.
AT Giuliante. A high speed directional comparison protection for EHV transmission lines. 18th Annual Western Protective Relay Conference, 1991.
SL Goldsborough et al. A new carrier relaying system. AIEE Trans 63:568–572, 1944.
GS Hope et al. Ultra high speed directional comparison relay for EHV/UHV lines. Trans CEA E&O Div, 26(Part 4):paper no. 87-sp-168, 1987.
IEEE Power System Relaying Committee. Fiber optic channels for protective relaying. IEEE Trans Power Delivery 4(1):165–176, 1989.
AL McConnel et al. Phase comparison carrier—current relaying. AIEE Trans 64: 825–832, 1945.
Power System Relaying Committee. Evaluation of transfer-trip relaying using power-line carrier. AIEE (PAS), 81:250–255, 1962.
NO Rice et al. A phase-comparison carrier—current-relaying system for broader applications. AIEE Trans, 71(Part III):246–249, 1952.
WJS Rogers. Optical fiber signaling for protection purposes. 4th International Conference on Developments in Power System Protection, IEEE Pub No 302, 1989, pp 130–134.
RGR Sanderson. Improved directional comparison algorithm for protection of multiterminal lines. 5th International Conference on Developments in Power System Protection, IEE Pub No 368, 1993, pp 153–156.
HT Seeley et al. All electronic one cycle carrier relaying system. AIEE (PAS), 73:161–195, 1954.

Chapter 6

JE Clem. Application of capacitance potential devices. AIEE Trans 48:1–8, 1939.
C Concordia et al. Transient characteristics of current transformers during faults. AIEE Trans 61:280–285, 1942.
A Cruden et al. Current measurement device based on the Faraday effect. 5th International Conference on Developments in Power System Protection, IEE Pub No 368, pp 69–72, 1993.
G Dalke. Confused current transformers. Texas A&M Protective Relaying Conference, pp 15–17, 1991.
EL Harder. Transient and steady state performance of potential devices. AIEE Trans 59:91–99, 1940.

IEEE Power System Relaying Committee. Transient response of current transformers. IEEE Trans (PAS), 96:1809–1811, 1977.
IEEE Power System Relaying Committee Report. Current transformers with secondary current rating lower than 5 A. IEEE Trans Power Delivery 3(2):501–506, 1988.
PG McLaren. Improved simulation models for current and voltage transformers in relay studies. IEEE Trans Power Delivery 7(1):152–159, 1992.
AT Sinks. Computation of accuracy of current transformers. AIEE Trans 59:663–668, 1940.
WK Sonnemann et al. Current transformers and relays for high speed differential protection. AIEE Trans 59:481–488, 1940.
A Sweetane. Transient response characteristics of potential devices. IEEE (PAS), 89:1989–1997, 1970.
EC Wentz. A simple method for determining of ratio error and phase angle error in current transformers. AIEE Trans 60:949–954, 1941.

Chapter 7

AIEE Relay Subcommittee. Relay protection of power transformers. AIEE Trans 66:911–916, 1947.
AJ McConnel. A generator differential relay. AIEE Trans 62:11–13, 1943.
HT Seeley. Effect of residual magnetism on differential current relays. Trans AIEE 2:164–169, 1943.

Chapter 8

JE Barkle et al. Protection of generators against unbalanced currents. AIEE Trans 72(Part III):282–285, 1953.
JE Barkle et al. Detection of grounds in generator field winding. AIEE Trans 74(Part III):467–470, 1955.
G Benmouyal et al. Field experience with digital relay for synchronous generators. IEEE Trans Power Delivery 7(4):1984–1992, 1992.
ETB Gross. Sensitive ground relaying of AC generators. AIEE Trans (Part III):539–541, 1952.
IEEE Power System Relaying Committee. Summary of the guide for AC generator protection. IEEE Trans Power Delivery 4(2):957–964, 1989.
CR Mason. A new loss of excitation relay for synchronous generators. AIEE Trans 68:1240–1245, 1949.
WC Morris et al. A negative-phase-sequence-overcurrent relay for generator protection. AIEE Trans 72(Part III):615–618, 1953.
MS Sachdev et al. A digital algorithm for detecting internal faults in synchronous generators. Trans CEA E&O Div 32, paper no 93-sp-103, 1993.
W Weijian et al. New developments of third harmonic ground fault protection schemes for turbine-generator stator windings. International Conference on

Developments in Power System Protection, IEEE Pub No 302, pp 250–253, 1989.
G Ziegler. Developments in generator protection—design and application aspects of a numerical relay range. 5th International Conference on Developments in Power System Protection, IEE Pub No 368, pp 111–114, 1993.

Chapter 9

AIEE Committee. Report on transformer magnetizing inrush currents and its influence on relaying and air switch operation. Trans IEEE 70(Part II):1730, 1951.
AIEE Relay Subcommittee. Relay protection of power transformers. AIEE Trans 1947.
K Anatha Raman. PC based IIR filter algorithm for transformer relaying. EPSR 123–127, 1993.
HL Cole et al. A sudden gas pressure relay for transformer protection. AIEE Trans 72(Part III):480–483, 1953.
AT Giuliante et al. Advances in the design of differential protection of power transformers. 45th Annual Texas A&M Protective Relay Conference, 1992.
CD Hayward. Harmonic current restrained relays for transformer differential protection. AIEE Trans 1941.
B Jeyasurya et al. A state-of-art review of transformer protection algorithm. IEEE Trans (PWDR) 3(2):531–541, 1988.
JT Madill. Typical transformer faults and gas detector relay protection. AIEE Trans 66:1052–1060, 1947.
CA Mathews. An improved transformer differential relay. AIEE Trans 73(Part III):645–649, 1954.
P Mudditt et al. Developments in transformer protection. 4th International Conference on Developments in Power System Protection. IEE Pub No 302, pp 61–65, 1989.
GK Rockfeller et al. Magnetizing inrush phenomena in transformer banks. AIEE Trans (PAS) 77:884, 1958.
MS Sachdev et al. On-line identification of magnetizing inrush and internal faults in three-phase transformers. IEEE Trans (PWDR) 7(4):1885–1891, 1992.
G Stranne. Numerical protection systems for generators and generator transformer units. ABB Rev 27–38, 1993.

Chapter 10

AIEE Relay Subcommittee. Bus protection. AIEE Trans 58:206–211, 1939.
JR Linders et al. A half cycle bus differential relay and its applications. IEEE Trans (PAS) 1970.
JB Royale et al. Low impedance biased differential busbar protection for application to busbars of widely differing configuration. 4th International Conference on Developments in Power System Protection, IEE Pub 302, pp 40–44, 1989.
J Rushton. Busbar protection. Electrical Rev 1156–1159, 1957.

WK Sonnemann et al. Linear couplers for bus protection. AIEE Trans 61:241–248, 1942.

Chapter 11

J Cremer. Aspects of modern secondary relay testing equipment. 4th International Conference on Development in Power Systems Protection, IEE Pub 302, pp 225–229, 1989.
HW Dommel. A software-based EMTP real-time simulator. CEA E&O Div 32, Paper 93-SP-105, 1993.
DB Fakruddin. A new PC based package for protective relay testing and relaying studies. Recent Trends in Applied Systems Research, 1994.
GD Rockfeller et al. Differential relay testing program using EMTP simulations. 46th Georgia Tech Protection Relaying Conference, 1992.
P Todd. Preventive maintenance can reduce outages. Electric Power Light 66(2):20, 1988.
AC Webb. Computer generation of test quantities for testing protective relays. 4th International Conference on Developments in Power System Protection. IEE Pub No 302, pp 30–34, 1989.

Chapter 12

DWE Blatt. New method for monitoring and protection of high voltage switchyards. IEE Proc-C 138(3):228–232, 1991.
SH Horowitz et al. Adaptive transmission relaying. IEEE Trans (PWDR) 3(4): 1436–1445, 1988.
AT Johns et al. The application of neural network techniques to adaptive autoreclosure in protection equipment. 5th International Conference in Developments in Power System Protection, IEE Pub No 368, pp 161–164, 1993.
LR Johnson. Pyramid concept. Trans CEA E&O Div 32, Paper No 93-SP-82, 1993.
WA Mittelstadt. A new out-of-step relay with rate of change of apparent resistance augmentation. IEEE Trans (PAS) 102(3):631–639, 1983.
YG Paithankar. Fast (1-shift) orthogonal functions for extraction of the fundamental frequency component for computer relaying. EPSR 14:233–236, 1988.
YG Paithankar et al. A new set of orthogonal functions with aperidic sampling for protective relays. National Seminar on Power Scenario in India, paper D-36, 1995.
YG Paithankar. Some applications of multi input–multi output pulse sequential circuits. Under Scrutiny 1995.
YG Paithankar et al. Distance relays for EHV lines augmented by impedance derivatives. 9th National Power Systems Conference NPSC-96, vol II, pp 573–577, 11T Kanpur, India, 1996.
YG Paithankar. Ultra high-speed relay for EHV/UHV lines employing traveling waves. NPSC, Paper S4.5.1, Hyderabad, India.

Bibliography

AG Phadke et al. Adaptive out-of-step relaying using phasor measurement techniques. IEEE Comput Appl Power 6(4):12–17, 1993.
JB Scarborough. Numerical Mathematical Analysis. Johns Hopkins Press, 1950, pp 152–165.

Chapter 13

Asea Brown Bovri. ABB network control and protection. Pub No 1 MBD-YN, 1994.
CIGRE. Final Report on computer based protection and digital techniques in substations, CIGRE CE/SC 34 GT/WG 02, 1985.
CIGRE. An international survey of the present status and perspective of expert systems in power system analysis and techniques. CIGRE, CS/SC 38 GT/WG 02 TF 07, 1988.
CIGRE. Digital protection techniques and substation functions. CIGRE, CE/SC 34 GT/WG 01, 1989.
GEC Alsthom. Integrated digital substation control system. Pub No PSCN 3020.
IEEE Power System Relaying Committee Working Group. Potential applications of expert systems to power system protection. IEEE Trans Power Delivery 9(2): 720–728, 1994.
Proceedings of the Third Symposium on Expert Systems Application to Power Systems, Tokyo, 1991.
Don Russel et al. Power System Control and Protection. New York: Academic Press, 1978.
Siemens literature on numerical protection equipment, Type LSA 678.
SJ Steel et al. PROSET: An expert system for protective relay setting. Proceedings of the IFAC Symposium on Power Systems and Power Plant Control, Korea, 1989, pp 1003–1007.
H Ungrad et al. Protection Techniques in Electrical Energy Systems. New York: Marcel Dekker, 1995.

Index

acceleration, 131
accuracy of relays, 34
adaptive relays, 344
aging of insulation, 1
algorithm, 347
alternator protection, 233
amplitude
  check, 250
  comparator, 97
analog-to-digital converter, 336
arcing, 85
attracted armature, 77
autoreclosure, 182

backup protection (remote and local), 11
balanced beam, 89
biased differential, 228
blinders, 175
Buchholtz, 266
burden (CT and PT), 201
busbar protection, 269

carrier, 185
carrier-aided distance scheme, 186
carrier unit schemes, 191
causes of faults, 1
check zone (busbar), 279

circuit breaker (CB), 4, 12
  duty, 20
coincidence period, 101
compensation of ground-fault relay, 142
coordination, 34
core balance, 236
cosine comparator, 102
current-graded OC relays, 31
current transformer (CT), 200

dc offset, 153
decelerating power, 131
definite time overcurrent (DTOC) relay, 25
deionization, 2, 326
deionizing times, 183
differential relay, 217
directional relay, 50
discrete Fourier transform, 122
discrimination, 12
distance protection, 75
double-end feed, 79
duality, 105
dynamic relays, 313
dynamic testing, 299

earth fault protection, 2
excitation characteristics, 208

379

export of power, 79
external fault, 219

fault, 1
   characteristics of line
      double-end feed
         (prefault export and import), 79, 80
      single-end feed, 78
   data, 68
   detector (distance relays), 147
   statistics, 3, 181
field
   failure, 241
   suppression, 234
filters, 336
flashover, 2
forward fault, 52
Fourier, 122
fuse, 36

generator protection, 233
generator-transformer unit, 268
governor, 243
ground fault, 2, 141
grounding, 233

harmonic relays, 307
harmonic restrained differential relay, 263
harmonics, 121
high-impedance fault, 67
high-set OC relay, 38

import of power, 80
induction cup unit, 59
instability, 20, 170
interconnector, 19
interfacing, 121

internal fault, 220
intertipping w/o carrier, 320
interturn protection, 236
inverse definite minimum line (IDMT) relay, 29, 31

knee-point voltage, 208

lightning, 1
lockout of CB, 182, 313
longest line, 159
loss of excitation, 241

magnetizing inrush, 259
maintenance of relays, 289
malfunction, 168
maximum torque angle, 60
measurement of R, X and Z, 128
memory (tuned circuit), 93
Merz-Price scheme, 219
mesh network, 14
metallic fault, 85
microprocessor-based relays
   directional, 64
   distance (conventional), 120
   overcurrent, 47
   quadrilateral, 130, 358
modeling, 138
multishot reclosure, 20

nature of faults, 1
negative-sequence relay, 237
neural, 321
noise, 314
noise-based relays, 314
nonswitched distance scheme, 145

Index

offset, 153
orthogonal functions, 122
outage, 172, 174
out-of-step
   blocking, 172
   tripping, 174
overcurrent (OC) relays
   DTOC, 25, 45
   high set, 38
   IDMT, 29, 45
overfluxing, 264
overheating/overspeed/overload, 243
overlapping of protective zones, 15
overreach (transient)
   distance relays, 153
   OC relays, 40
overshoot, 33

parallel feeders, 14
percentage differential relay, 228
permanent fault, 2
phase
   check, 250
   comparator, 98
   comparison carrier, 193
phase-fault relays
   distance, 137
   overcurrent, 68
plug setting, 33
polarization, 57
polarized relay, 263
polyphase relay, 150
potential transformer (PT), 210
power line carrier, 185
power swing (stable and unstable), 130
pressure relief relay, 266
primary protection, 10
programming, 344
protective zone, 16

quadrilateral distance relay
   multi-input amplitude comparator, 119
   multi-input phase comparator, 113

ratio correction factor, 207
reach, 76
reactance relay
   amplitude comparator, 110
   phase comparator, 109
reclosure, 20
reliability, 20
remote backup, 12
replica burden, 210
residual flux, 260
restricted earth fault relay, 257
resynchronization, 174
reverse power relay, 243
ring main protection, 14
ring modulator, 63
rotor faults, 239

sampling, 122
saturation, 204, 207
Schmitt trigger, 102
security, 21
selectivity, 12
sensitivity, 4
shortest line, 162
short-time current rating, 37
single-end-feed lines, 78
spill current, 224
split-phase protection (general), 236
stability, 152
stability ratio, 220, 278
static relays
   distance, 96
   overcurrent, 45
stator protection, 233

steady-state testing, 289
summation transformer, 196
supervisory relay (busbar), 279
surges, 1
switched distance scheme, 147
swiveling characteristics, 167

testing relays, 289
three-stepped distance scheme, 153
thyristors, 96
tie line, 20
time-current characteristics, 31, 45
time delay, 25
time-graded OC relays, 34
transformer
    connections, 247
    protection, 247
transients, 259
trap, 185

unbalance (general), 237
underreaching, 151
unit protection, 217
unsaturated current transformer, 209
unsuccessful reclosure, 20

variable-slope differential relay, 230
vibrations, 29

Walsh orthogonal functions, 124
waveform shaping, 302
weighting coefficients, 123, 124

Zener diode, 46
zero sequence compensation, 143
zones of protection, 8